Energy Science, Engineering and Technology

Emissions Control in Electricity Generation

ENERGY SCIENCE, ENGINEERING AND TECHNOLOGY

Additional books in this series can be found on Nova's website under the Series tab.

Additional E-books in this series can be found on Nova's website under the E-book tab.

ENERGY POLICIES, POLITICS AND PRICES

Additional books in this series can be found on Nova's website under the Series tab.

Additional E-books in this series can be found on Nova's website under the E-book tab.

ENERGY SCIENCE, ENGINEERING AND TECHNOLOGY

EMISSIONS CONTROL IN ELECTRICITY GENERATION

JACKSON L. WALKER
AND
ALLEN HERNANDEZ
EDITORS

nova publishers
New York

For permission to use material from this book please contact us:
Telephone 631-231-7269; Fax 631-231-8175
Web Site: http://www.novapublishers.com

NOTICE TO THE READER

Additional color graphics may be available in the e-book version of this book.

Library of Congress Cataloging-in-Publication Data

ISBN: 978-1-62257-830-6

Published by Nova Science Publishers, Inc. † New York

CONTENTS

PREFACE

Federal policymakers have long been concerned about the impact of manmade carbon dioxide (CO2) emissions on global climate change. To address these concerns, Congress has been debating a range of potential initiatives for reducing atmostpheric CO2 from U.S. sources. Legislative proposals would seek to limit U.S. CO2 emissions to specific (historical) levels through emissions caps, carbon taxes, or other regulatory mechanisms. Many of these proposals dictate or anticipate a declining long-term trajectory for annual U.S. carbon emissions. The most prominent CO2 proposals seek reductions of nationwide CO2 emissions to 1990 levels or lower by 2030. This book examines key uncertainties associated with CO2 emissions abatement measures, with a focus on available and emerging technologies, and new source review permits.

Chapter 1 – Congress has been debating a range of potential initiatives for reducing atmospheric CO_2 from U.S. sources. Legislative proposals would seek to limit U.S. CO_2 emissions to historical levels through emissions caps, carbon taxes, or other mechanisms. In the 110th Congress, the most prominent CO_2 proposals sought reductions of nationwide CO_2 emissions to 1990 levels or lower by 2030. President Obama has proposed cutting carbon CO_2 emissions to 1990 levels by 2020, and by an additional 80% by 2050.

A fundamental question arising from carbon control proposals is how the CO_2 reduction targets can be achieved in the electricity industry, which is responsible for nearly 40% of U.S. CO_2 emissions. It appears from the policy research and technical studies that substantially reducing CO_2 emissions in the U.S. electricity sector over the next few decades would likely require every key carbon mitigation measure at the nation's disposal. However, it is also clear that significant uncertainty exists about the potential of individual measures to achieve their hoped-for carbon impact:

- **Energy efficiency**—Can the United States overcome socioeconomic barriers to achieve four times more potential savings than ever before?
- **Renewable energy**—Will there be enough transmission for wind power? Is there enough land to grow the needed biomass?
- **Nuclear power**—Could the United States build new plants fast enough to matter?
- **Advanced coal power**—Will banks fund them and regulators approve them?
- **Carbon capture and sequestration**—Will the technology be commercially deployable in 10 years, 25 years, or never?
- **Plug-in hybrid electric vehicles**—How much "low carbon" electricity would be available to charge their batteries?
- **Distributed energy resources**—Would carbon costs change distributed energy economics enough to spur deployment?

As the nation's CO_2 mitigation policies develop, the inherent uncertainty associated with specific carbon measures may be a critical concern. Commitments to specific carbon emissions targets over time, or to a specific schedule of carbon costs (whatever form they may take) may be greatly affected by the success of the underlying measures relied upon to achieve them. Notwithstanding the best efforts of federal policy makers, it is possible that, given the uncertainties each faces, few if any of the major measures proposed to moderate U.S. carbon emissions will achieve their anticipated impacts in a 20-year time frame. As Congress considers implementing CO_2 policies, keeping a close eye on the technology and market developments associated with every key measure could be a priority. Balancing responses to energy market volatility and unexpected structural changes against the need for a predictability in R&D and private capital investment may be essential to maintaining the nation on course to meaningful atmospheric CO_2 reduction.

Chapter 2 – This document is one of several white papers that summarize readily available information on control techniques and measures to mitigate greenhouse gas (GHG) emissions from specific industrial sectors. These white papers are solely intended to provide basic information on GHG control technologies and reduction measures in order to assist States and local air pollution control agencies, tribal authorities, and regulated entities in implementing technologies or measures to reduce GHGs under the Clean Air Act, particularly in permitting under the prevention of significant deterioration (PSD) program and the assessment of best available control technology (BACT). These white papers do not set policy, standards or otherwise establish any binding requirements; such requirements are contained in the applicable EPA regulations and approved state implementation plans.

This document provides information on control techniques and measures that are available to mitigate GHG emissions from the coal-fired electric generating sector at this time. The primary GHG emitted by the coal-fired electric generation industry is carbon dioxide (CO_2), and the control technologies and measures presented in this document focus on this pollutant. While a large number of available technologies are discussed here, this paper does not necessarily represent all potentially available technologies or measures that that may be considered for any given source for the purposes of reducing its GHG emissions. For example, controls that are applied to other industrial source categories with exhaust streams similar to the cement manufacturing sector may be available through "technology transfer" or new technologies may be developed for use in this sector.

The information presented in this document does not represent U.S. EPA endorsement of any particular control strategy. As such, it should not be construed as EPA approval of a particular control technology or measure, or of the emissions reductions that could be achieved by a particular unit or source under review.

Chapter 3 – Older electricity generating units—those that began operating in or before 1978—provided 45 percent of electricity from fossil fuel units in 2010 but produced a disproportionate share of emissions, both in aggregate and per unit of electricity generated. Overall, in 2010 older units contributed 75 percent of sulfur dioxide emissions, 64 percent of nitrogen oxides emissions, and 54 percent of carbon dioxide emissions from fossil fuel units. For each unit of electricity generated, older units collectively emitted about 3.6 times as much sulfur dioxide, 2.1 times as much nitrogen oxides, and 1.3 times as much carbon dioxide as newer units. The difference in emissions between older units and their newer counterparts may be attributed to a number of factors. First, 93 percent of the electricity produced by older fossil fuel units in 2010 was generated by coal-fired units. Compared with natural gas units, coal-fired units produced over 90 times as much sulfur dioxide, twice as much carbon dioxide and over five times as much nitrogen oxides per unit of electricity, largely because coal contains more sulfur and carbon than natural gas. Second, fewer older units have installed emissions controls, which reduce emissions by limiting their formation or capturing them after they are formed. Among coal-fired units—which produce nearly all sulfur dioxide emissions from electric power generation— approximately 26 percent of older units used controls for sulfur dioxide, compared with 63 percent of newer units. Controls for nitrogen oxide emissions were more common among all types of fossil fuel units, but these controls vary widely in their effectiveness. Among older units, 14 percent had installed selective catalytic reduction (SCR) equipment, the type of control capable of reducing the greatest amount of nitrogen oxides emissions, compared with 33 percent of newer units. In addition, approximately 38 percent of older units did not have any controls for nitrogen oxides, compared with 6 percent of newer units. Third, lower emissions among newer units may be attributable in part to improvements in the efficiency with which newer units convert fuel into electricity. Nonetheless, older units remain an important part of the electricity generating sector, particularly in certain regions of the United States.

Chapter 4 – Electricity generating units that burn fossil fuels supply most of the nation's electricity and are major sources of air pollution. Under the Clean Air Act, such units are subject to NSR, a permitting process that applies to (1) units built after August 7, 1977, and (2) existing units that undertake a major modification. Owners of such units must obtain from the appropriate permitting agency a preconstruction permit that sets emission limits and requires the use of certain pollution control technologies. EPA oversees states' implementation of NSR, including reviewing and commenting on draft permits issued by state and local permitting agencies. GAO was asked to examine (1) what information EPA maintains on NSR permits issued to fossil fuel electricity generating units; (2) challenges, if any, that EPA, state, and local agencies face in ensuring compliance with requirements to obtain NSR permits; and (3) what available data show about compliance with requirements to obtain NSR permits. GAO reviewed relevant documentation and interviewed EPA, state, and local officials, as well as representatives from industry, research, and environmental groups.

In: Emissions Control in Electricity Generation ISBN: 978-1-62257-830-6
Editors: J. L. Walker, A. Hernandez

Chapter 1

CARBON CONTROL IN THE U.S. ELECTRICITY SECTOR: KEY IMPLEMENTATION UNCERTAINTIES*

Paul W. Parfomak

SUMMARY

Congress has been debating a range of potential initiatives for reducing atmospheric CO_2 from U.S. sources. Legislative proposals would seek to limit U.S. CO_2 emissions to historical levels through emissions caps, carbon taxes, or other mechanisms. In the 110th Congress, the most prominent CO_2 proposals sought reductions of nationwide CO_2 emissions to 1990 levels or lower by 2030. President Obama has proposed cutting carbon CO_2 emissions to 1990 levels by 2020, and by an additional 80% by 2050.

A fundamental question arising from carbon control proposals is how the CO_2 reduction targets can be achieved in the electricity industry, which is responsible for nearly 40% of U.S. CO_2 emissions. It appears from the policy research and technical studies that substantially reducing CO_2 emissions in the U.S. electricity sector over the next few decades would likely require every key carbon mitigation measure at the nation's disposal. However, it is also clear that significant uncertainty exists about

---

* This is an edited, reformatted and augmented version of Congressional Research Service, Publication No. R40103, dated December 23, 2008.

the potential of individual measures to achieve their hoped-for carbon impact:

- **Energy efficiency**—Can the United States overcome socioeconomic barriers to achieve four times more potential savings than ever before?
- **Renewable energy**—Will there be enough transmission for wind power? Is there enough land to grow the needed biomass?
- **Nuclear power**—Could the United States build new plants fast enough to matter?
- **Advanced coal power**—Will banks fund them and regulators approve them?
- **Carbon capture and sequestration**—Will the technology be commercially deployable in 10 years, 25 years, or never?
- **Plug-in hybrid electric vehicles**—How much "low carbon" electricity would be available to charge their batteries?
- **Distributed energy resources**—Would carbon costs change distributed energy economics enough to spur deployment?

As the nation's CO_2 mitigation policies develop, the inherent uncertainty associated with specific carbon measures may be a critical concern. Commitments to specific carbon emissions targets over time, or to a specific schedule of carbon costs (whatever form they may take) may be greatly affected by the success of the underlying measures relied upon to achieve them. Notwithstanding the best efforts of federal policy makers, it is possible that, given the uncertainties each faces, few if any of the major measures proposed to moderate U.S. carbon emissions will achieve their anticipated impacts in a 20-year time frame. As Congress considers implementing CO_2 policies, keeping a close eye on the technology and market developments associated with every key measure could be a priority. Balancing responses to energy market volatility and unexpected structural changes against the need for a predictability in R&D and private capital investment may be essential to maintaining the nation on course to meaningful atmospheric CO_2 reduction.

INTRODUCTION

Federal policymakers have long been concerned about the impact of manmade carbon dioxide (CO_2) emissions on global climate change. To address these concerns, Congress has been debating a range of potential initiatives for reducing atmospheric CO_2 from U.S. sources. Legislative proposals would seek to limit U.S. CO_2 emissions to specific (historical) levels through emissions caps, carbon taxes, or other regulatory mechanisms. Many of these proposals dictate or anticipate a declining long-term trajectory for annual U.S. carbon emissions. In the 110th Congress, the most prominent CO_2 proposals sought reductions of nationwide CO_2 emissions to 1990 levels or lower by 2030.[1] President Obama has proposed carbon reduction targets as well, intending to cut CO_2 emissions to 1990 levels by 2020, and by an additional 80% by 2050.[2]

A fundamental policy question which arises from carbon control proposals is how the CO_2 reduction targets can be achieved. Numerous analysts have been examining this question and identifying specific measures to reach particular targets—especially in the electricity industry, which is responsible for nearly 40% of U.S. CO_2 emissions. In the electricity sector, these measures typically include some combination of energy efficiency, renewable energy, nuclear power, advanced fossil-fuel power generation, carbon capture and sequestration, plug-in hybrid electric vehicles, and distributed energy resources. **Figure 1** illustrates the CO_2 abatement potential of these measures in the electricity sector as estimated in a widely cited analysis by the Electric Power Research Institute (EPRI). In the EPRI example, overall CO_2 emissions associated with the electricity sector would be reduced below 1990 levels by 2030. Other studies have generated their own projections, with alternative targets and assumptions leading to distinct trajectories for CO_2 emissions and different contributions from the various abatement measures.[3]

Analyses like that in **Figure 1** are important for understanding the potential opportunities and limitations of specific technological measures which may be needed to meet CO_2 abatement goals. However, with their focus on technical potential, such studies often have difficulty conveying in a straightforward way the key infrastructural, environmental, regulatory, or operational uncertainties which might affect how much of that potential could practically be achieved.[4] These uncertainties, nonetheless, are of critical concern to legislators overseeing existing carbon-related programs or considering future CO_2 abatement policies. In addition to their technical aspects, Congress faces a need to gauge the overall viability of specific CO_2 abatement measures—what are their prospects for helping to achieve CO_2 targets in the electricity sector. Legislators also face a need to assess the time frame over which such measures could be expected to work, and how these measures may fit together to achieve overall CO_2 abatement goals under a national carbon policy.

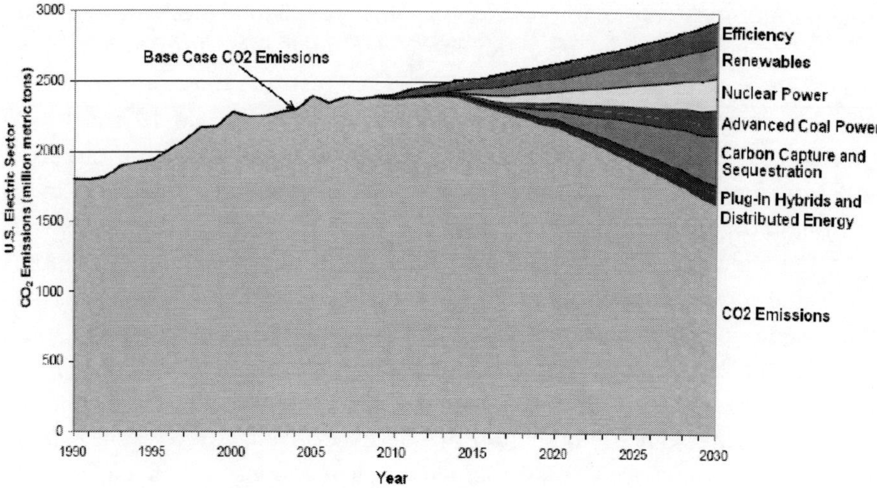

Source: Adapted from Barbara Tyran, Electric Power Research Institute, "The Power to Reduce CO_2 Emissions: The Full Portfolio," Slide presentation (May 15, 2008): 7. http://www.iea.org/ Textbase/work/2008/roadmap/2a_Tyran_EPRI%2 0Road maps.pdf.

Figure 1. Potential CO_2 Reductions in Electric Power.

This report examines key uncertainties associated with the CO_2 emissions abatement measures identified in **Figure 1**. The report briefly describes each measure and discusses expectations for its potential in the context of past experience, technical challenges, infrastructure requirements, or other factors which may inform performance expectations. For each measure, it identifies and discusses a critical uncertainty which may influence its overall viability. The report concludes with a discussion of the implications of these uncertainties in the context of the congressional carbon control debate.

ENERGY EFFICIENCY AND CONSERVATION

When an energy conversion device, such as a household appliance or automobile engine, undergoes a technical change that enables it to provide the same service (e.g, cooling, lighting, motor drive) while using less energy it is said to have increased "energy efficiency." The energy-saving result of the efficiency increase is often called "energy conservation." The energy efficiency of buildings can likewise be increased through the use of certain design changes such as more insulation, thermal windows, improved ventilation, and solar orientation.[5] Energy efficiency is often viewed as interchangeable with energy supply options like electric generation, oil, or natural gas. Energy efficiency can also reduce energy resource use and associated environmental impacts—like CO_2 emissions from power plants.

Electricity-Efficiency Potential

Baseline improvements in energy efficiency occur over time as an economic response to changes in energy prices, the availability of new technology, turnover in end-use equipment, and other factors. However, conservation studies since the OPEC oil embargoes of the 1970s have identified substantial potential for energy efficiency improvements above baseline levels. One seminal analysis in 1976 stated

> technical fixes in new buildings can save 50 percent or more in office buildings and 80 percent or more in some new houses.... [B]y 1990, improved design of new buildings and modification of old ones could save a third of our current total national energy use—and save money too.[6]

A 1981 study by the Solar Energy Research Institute[7] likewise found that "through energy efficiency, the U.S. can achieve a full-employment economy and increase worker productivity, while reducing national energy consumption by nearly 25 percent."[8] The study further concluded that "the consumption of electricity can be reduced to a point where, on a national basis, demands through the end of the [twentieth] century can be met with generating equipment now operating or in advanced stages of construction."[9]

More recent studies continue to identify significant electricity conservation potential. The "Five Lab Study" in 1997 estimated a technical electricity savings potential of approximately 23%, and a maximum "achievable" potential of 15% among residential and commercial buildings, assuming aggressive policies promoting conservation and a $50/metric tons (1993 dollars) carbon cost.[10] A 2004 meta-analysis by the American Council for an Energy-Efficient Economy of several regional studies reported a technical electricity conservation potential of 33%, and an achievable potential of 24% over a 5 to 15 year time horizon, depending upon the study.[11] The U.S. Department of State's 2006 *Climate Action Report* concludes that "by using commercially available, energy-efficient products, technologies, and best practices, many commercial buildings and homes could save up to 30 percent on energy bills."[12]

Impacts from Electricity-Efficiency Initiatives

Both federal and state agencies have implemented a multitude of electricity efficiency initiatives over the last 40 years to capture energy efficiency potential in the electricity sector. These initiatives have included appliance, equipment, and building efficiency standards; electric utility-administered conservation incentives;[13] consumer information campaigns; and other initiatives. Notwithstanding these efforts, the levels of incremental electricity conservation actually achieved since the 1970s have been more modest than the 25%-30% suggested in conservation potential studies. A 2004 analysis examining a comprehensive range of both federal and utility-sponsored conservation and energy efficiency programs (including federal efficiency standards) administered through 2000 concluded as follows:

> [P]rograms for which *ex post* quantitative estimates of energy savings exist are likely to have collectively saved up to 4.1 quads of electricity annually. These estimates typically reflect the cumulative effect of programs (e.g., all appliance efficiency standards, past and present) on annual energy consumption. This total energy savings represents about 6% of annual nontransportation energy consumption....[14]

A study of California's 2001 energy demand reduction initiative (promoted heavily as an emergency measure to avoid blackouts during the state's electricity crisis) reported 6% reduced electricity usage compared to the prior year, although only 25%-30% of this reduction was "attributable to savings from energy efficiency or onsite generation projects ... likely to persist for many years."[15] Consistent with these studies, a 2008 analysis by EPRI projected a "realistic" U.S. electricity savings potential of 7% beyond baseline levels which would occur without additional market intervention.[16]

Uncertainty about the Efficiency Opportunity

Taken together, the studies of technical conservation potential and actual conservation impacts suggest a perpetual opportunity for incremental electricity conservation on the order of 25% more than four times the savings such programs have actually realized. Moving beyond the 5% to 7% electricity savings range has been a persistent challenge to conservation proponents, primarily because of the diffuse nature of the efficiency opportunity and the economic complexity of decision making and capital investment by electricity consumers.

> Students of end-use markets have long been puzzled by the lack of adoption of ostensibly cost-effective energy efficiency technologies. A rich literature has developed around this question, and evidence for various barriers to adoption of efficiency technologies is widespread.[17]

Among the barriers to energy efficiency analysts have identified are–

- limited market availability of new efficiency measures,
- incomplete consumer information about efficiency options,
- insufficient capital for efficiency investments,
- fiscal or regulatory policies discouraging efficiency investments,
- builder focus on first costs vs. lifecycle costs,
- lack of consumer focus on energy costs relative to other costs, and
- energy prices not reflecting the full social costs of energy supply.[18]

Conservation advocates and federal policy makers have proposed a range of additional policy approaches to further overcome these barriers, but there is limited consensus on which policies would be effective and how much additional conservation they might achieve. As a study of conservation barriers from Lawrence Berkeley National Laboratory concluded–

> Although these rationales provide a basis for some type of intervention, we acknowledge that they do not justify any particular intervention.... [W]e suggest that differences of opinion about the appropriateness of public policies stem not from disputes about whether market barriers exist, but from different perceptions of the magnitude of the barriers and the efficacy and (possibly unintended) consequences of policies designed to overcome them....[19]

There have been numerous legislative proposals to promote electricity efficiency and conservation.[20] The key uncertainty faced by all of them, and future conservation proposals, is whether, as a whole, they may cost-effectively capture much more of the "latent" electricity conservation potential than such programs have done in the past. Carbon control studies which project electricity efficiency savings on the order of 5% to 10% over a 20-year time frame appear consistent with U.S. conservation program experience, and may be aided by any future costs of CO_2 emissions if they are reflected in electricity prices. Unlike large physical infrastructure, however, such as power plants or electric transmission lines, conservation impacts do not necessarily "scale up" to achieve greater impacts simply by increasing the size or funding of a given conservation program. Efficiency potential is also extremely diffuse—existing literally at the individual light socket level in nearly every household. Consequently, policy makers seeking large conservation impacts may need to try alternative or more aggressive policies (e.g., very strict building efficiency codes) with little track record upon which to base projections of future effectiveness.

RENEWABLE ENERGY

Renewable energy supplies in the electricity sector typically include the following types of power plants: geothermal, solar, wind, biomass, and municipal solid waste/landfill gas.[21] Many carbon control advocates and federal policy makers have high expectations for the potential of renewable generation to help reduce CO_2 emissions in the United States. The widely publicized Pickens Plan, for example, announced by T. Boone Pickens in July, 2008, envisions deploying enough wind generation in the Great Plains states to produce 20% of U.S. electricity by 2018.[22] The Obama-Biden presidential campaign similarly pledged to establish a federal Renewable Portfolio Standard requiring that 25 percent of electricity consumed in the U.S. be "derived from clean, sustainable energy sources, like solar, wind, and geothermal" by 2025.[23] Some groups have advocated even more aggressive targets for U.S. renewable power development.[24] Others are less optimistic. EPRI, for example, assumes renewable sources (excluding hydroelectric generation) could contribute 9% of electricity production in the electric power sector by 2030.[25]

According to the Energy Information Administration (EIA), biomass and wind generation are the two types of renewable power with the greatest overall economic potential, and therefore, the greatest potential to reduce CO_2 emissions.[26] Biomass and wind power are inherently different technologies, however, so they face distinct uncertainties related to their potential expansion under a national program of carbon control. They are discussed, in turn, in the following sections.

Wind Power

Wind power does not consume fuel and produces no CO_2, so it is an extremely attractive technology for CO_2 mitigation in the electricity sector.[27] Wind generation technology is also fairly mature. It has been deployed commercially throughout the United States—albeit with federal assistance in the form of renewable energy production tax credits and state renewable energy portfolio standards.[28] Wind plants accounted for just over 1% of U.S. electricity generation in 2008.[29]

A 1991 study of available U.S. wind resources by Pacific Northwest Laboratory concluded that "the wind electric potential that could be extracted with today's technology ... across the United States is equivalent to about 20% of the current U.S. electric consumption."[30] A 2003 analysis by the National Renewable Laboratory similarly concluded that "by 2050, wind could account for about 25 percent of all generation in the U.S."[31] Consistent with these technical assessments, the Department of Energy (DOE) recently examined the practical possibility of 20% wind power production in the United States by 2030 (a longer time frame than either the Pickens or ObamaBiden campaign proposals). The DOE report concluded that a "20% Wind Scenario in 2030, while ambitious, could be feasible."[32]

Transmission Requirements

Although studies have identified substantial untapped wind power potential for the United States, the rapid expansion of U.S. wind generation faces significant challenges. According to the DOE, these challenges are related to the integration of intermittent wind power into regional electricity control areas, cost reduction and efficiency improvement for wind turbine technology, and wind facility siting issues.[33] The principal challenge the DOE identifies, however—and, according to many experts, the principal uncertainty facing wind power—is "investment in the nation's transmission system."[34]

As **Figure 2** shows, the nation's most abundant wind resources tend to be located far from population centers where the electricity is needed. Consequently, wind generators require a robust transmission grid to move power to the market. But the U.S. transmission network is constrained, significantly limiting the availability of transmission capacity to new wind farms. Transmission owners, in agreement with the DOE, have pointed to transmission grid constraints as the single greatest impediment to aggressive wind power expansion.

> Although there is sufficient evidence showing that wind generation can be reliably integrated into the electricity system, ... obstacles to new generation sources continue to exist due to the lack of adequate transmission system access. The remoteness of wind sources, an underinvested transmission infrastructure, and lack of workable transmission investment policies all hinder the development of wind power in the US.[35]

The DOE study estimates that to achieve 20% wind energy in the United States would involve building "more than 12,000 miles of additional transmission, at a cost of approximately $20 billion in net present-value terms."[36] A similar conceptual transmission plan by the American Electric Power Company (AEP) to integrate 20% wind energy estimated that 19,000 miles of new 765 kV transmission lines would be needed, with a net present value of $26 billion.[37] While such plans put a needed focus on some of the specific requirements for a U.S. transmission upgrade, marshaling the level of investment to expand transmission capacity quickly could stress the supply and price of materials, labor, and other resources. Any wind power project requiring the construction of extensive new transmission infrastructure from remote to populated areas also could face concerted community opposition to the siting of those transmission lines. Public challenges to electric transmission projects have long been considered among the most serious and most intractable challenges in the U.S. energy sector.[38] With wind projects in mind, Congress included provisions increasing federal authority to approve interstate electric transmission projects in the Energy Policy Act of 2005 (P.L. 109-58 § 1221). Nonetheless, challenges continue to delay or prevent new transmission development in some regions.

Transmission Grid Uncertainty

If U.S. wind power is beginning to face transmission constraints at only 1% of total U.S. electricity production, analysts raise concerns about the practicality of transmitting 20 times that amount of wind power in the near term. In its 2007 wind program plans, for example, the DOE itself states that large wind power deployment projections "seem too good to be true."[39] A 2008 *New York Times* assessment makes the point more strongly: "Experts say that without a solution to the grid problem, effective use of wind power on a wide scale is likely to remain a dream."[40] The key uncertainty for wind power then, is whether the electric grid, after decades of under-investment, will expand sufficiently to support a rapid expansion of wind power.

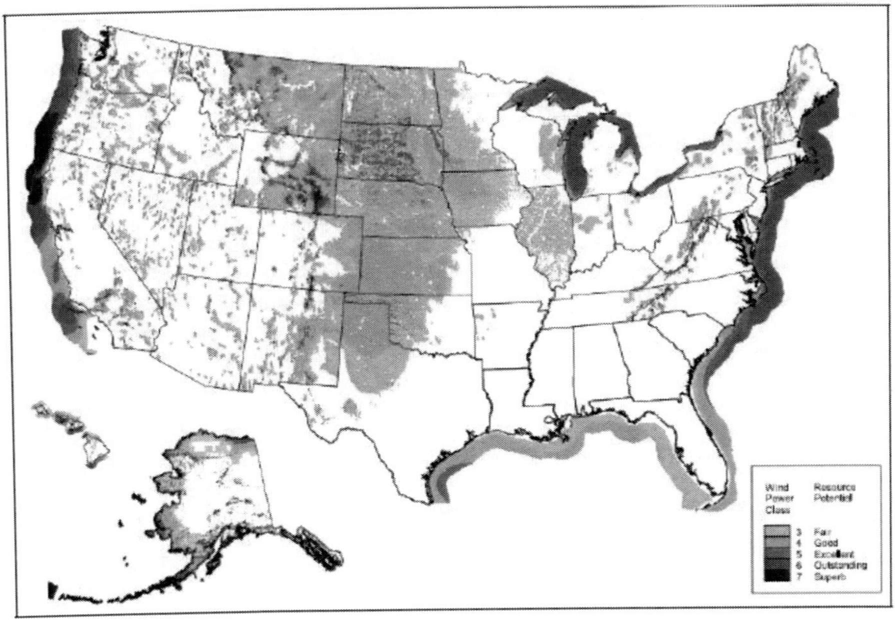

Source: National Renewable Energy Laboratory (2008).

Figure 2. Wind Power Resources in the United States.

Biomass Power Generation

Biomass power plants are combustion power plants that effectively recycle the CO_2 they emit through carbon sequestration in the crops grown (continuously) for fuel. Crops take up carbon dioxide from the air via photosynthesis as they grow and release it to the air when they are burned, so they cause no net increase in atmospheric CO_2.[41] Currently, most biomass power plants are fueled with waste materials from farming, forestry, and manufacturing (e.g., paper mill byproducts), although future expansion of biomass generation is expected to rely increasingly on dedicated fuel crops, such as poplar and switchgrass. Biomass power plants accounted for 1.3% of U.S. electricity generation in 2008.[42]

Biomass Fuel Supply

The principal factor which constrains the potential expansion of biomass power in the United States is the availability of biomass fuel. Biomass crops dedicated for power generation require land to grow—potentially in competition with food crops, lumber, and other traditional crops. Up to a point, such competition may not be a significant barrier to growth, since biomass power producers can use more waste from existing agricultural production (e.g., corn stalks) or grow fuel crops on U.S. lands not currently in agricultural production. A 2002 analysis by the Energy Information Administration concluded that U.S. biomass power generation capacity could increase by a factor of ten (from approximately 7 GW to 70 GW) through 2020 and not conflict with land requirements for existing crop production.[43] A 2005 analysis by the Department of Agriculture similarly concluded that forestland and cropland had the potential to support a sevenfold increase in the amount of biomass consumed for "bioenergy and biobased products."[44]

Biomass Fuel Uncertainty

While studies like those above seem to support the potential expansion of biomass power production, others contradict them, raising critical questions about the limits of biomass fuel supply in the United States. As a 2008 RAND report states, "the cost and supply of future biomass feedstocks are highly uncertain factors."[45] This uncertainty is exacerbated by the possible expansion of biomass demand for transportation fuels such as ethanol and biodiesel—also motivated by CO_2 abatement objectives, but in the transportation sector. To the extent that biomass for power and biomass for liquid fuels are pursued aggressively and concurrently, overall competition for U.S. agricultural resources may become a serious concern.

> The significant resulting increase in biomass usage would require harvesting various energy crops at a scale that vastly exceeds current practice. Greatly increased biomass production could be accompanied by adverse environmental and economic impacts due to land conversion.[46]

Adverse impacts limiting biomass production for electric power could include increases in food prices. Some analysts, for example, have argued that corn price increases could be partly linked to the diversion of U.S. corn crops from food and feedstock supply to fuel ethanol production.[47] Biomass crops could also compete for land with forest carbon sequestration, the latter likely to increase in importance as a source of valuable CO_2 emissions offsets under future carbon control policies.[48] Some analysts have suggested that biomass fuel or food crops could be imported to alleviate domestic land constraints, but such imports would raise other concerns about U.S. energy and food supply independence. A 2007 MIT study concluded,

> If we restrict USA biofuels to those produced domestically, as much as 500 million acres of land would be required in the USA for biofuels production.... The result would be that the USA would need to become a substantial agricultural importer. This suggests that the idea that biomass energy represents a significant domestic energy resource in the USA is misplaced.[49]

Carbon release from previously untilled lands and from agricultural processing are other factors which may reduce the life cycle CO_2 benefits, and therefore the economic potential, of rapid biomass crop expansion.

NUCLEAR POWER GENERATION

Nuclear power has been a significant part of the U.S. electricity sector for decades. The U.S. nuclear power industry currently comprises 104 licensed reactors at 65 plant sites in 31 states and generates about 20% of the nation's electricity.[50] Although no nuclear power plants have been ordered in the United States since 1978, and more than 100 reactors (all ordered after 1973) have been canceled, nuclear power is receiving renewed interest, prompted by a spike in fossil fuel prices, new federal subsidies and incentives—and possible CO_2 mitigation policies. As of September 30, 2008, the Nuclear Regulatory Commission had received, or anticipated receiving, license applications for 20 new nuclear reactor projects comprising 30 reactors in total.[51]

Because nuclear power generation is an established technology and produces virtually no direct CO_2 emissions, an expansion of U.S. nuclear power is viewed by many as essential for reaching long term CO_2 mitigation goals.[52] For example, in recent remarks before the Governor's Global Climate Summit, President-elect Obama stated that "we will tap nuclear power" as one way to help "build a clean energy future."[53] Likewise the Intergovernmental Panel on Climate Change *Fourth Assessment Report* states that "[n]uclear energy ... could make an increasing contribution to carbon free electricity and heat in the future."[54] EPRI proposes a 64% net increase in U.S. nuclear generating capacity, over four times the capacity addition in EIA's reference forecast, by 2030.[55] A 2003 study by MIT postulates a 300% net increase in U.S. nuclear power capacity by 2050.[56]

Nuclear Power Construction Uncertainty

Although nuclear energy proponents have high hopes for growth in this sector, not all groups concerned about climate change favor nuclear power as a key element of U.S. carbon control. The Natural Resources Defense Council, for example, argues that nuclear power may be too costly relative to clean energy alternatives, and may present unacceptable risks of nuclear proliferation and environmental damage from radioactive waste.[57] Other analysts raise questions about longterm nuclear fuel supply constraints, plant safety and security, and public acceptance.[58] These are all vital questions, any one of which could influence the viability of a U.S. nuclear resurgence. Perhaps a more fundamental uncertainty, however, even if these questions were resolved, is whether nuclear power plants could be constructed quickly enough to significantly reduce U.S. carbon emissions in a 20-year (or longer) time frame.

Uncertainty arises about the possible pace of U.S. nuclear expansion because nuclear power plants are large, complex, and must go through a lengthy and rigorous siting approval process. Furthermore, global capability to construct nuclear plants has diminished since the 1980s. A 2001 DOE analysis concluded that, because of limited growth in the nuclear sector over many years, there had been a "gradual erosion" in important nuclear infrastructure elements, such as qualified personnel in nuclear operations, qualified suppliers of nuclear equipment, and contractors with the necessary skills for nuclear design, engineering, and construction.[59] Little has changed since that report was released. Nuclear plant construction in the United States remains constrained by time, access to critical construction resources, and the availability of qualified engineering and construction firms.

Given the decline in U.S. nuclear infrastructure, there is ongoing debate among nuclear analysts about the prospect for rapid nuclear power expansion. A 2008 analysis by the Organisation for Economic Co-operation and Development (OECD) concluded that, based on historical experience from the 1970s and 1980s, and more recent growth in the world economy, "the capability could be rebuilt to construct 35-60 1000 GWe reactors per year" worldwide by 2030, and "[b]y 2050, capability could grow to 70-120 reactors per year."[60] Roughly consistent with this assessment, a 2008 study by the DOE's Nuclear Energy Advisory Committee concluded that, with additional infrastructure, it would be "plausible," albeit "a major challenge," to build 30 new U.S. reactors by 2030.[61] The latter rate of construction is substantially lower than that during the peak period of U.S. nuclear development, from 1963 to 1985, during which time 77 nuclear reactors were ordered, constructed, and began commercial operation.[62]

Others are more pessimistic about the potential for a rapid expansion of nuclear power in the United States. For example, one analyst concludes that "contrary to the public's perception and the industry's efforts, nuclear power will continue [a] long-term decline rather than move toward a flourishing future revival."[63] The debate is complicated by the need to replace U.S. nuclear plant capacity retiring in the near future simply to maintain nuclear power's current electricity supply contribution. Referring specifically to such construction limitations, another expert concludes that "nuclear energy will remain an option among efforts to control climate change, but given the maximum rate at which new reactors can be built, much new construction will simply offset the retirement of nuclear reactors built decades ago."[64] Observing that "the nuclear capability of the U.S. has atrophied in the 30 years since the last nuclear plant was ordered," Secretary of Energy Samuel Bodman has remarked more bluntly: "let's not kid ourselves about the challenges here." [65] In its 2008 report, the National Intelligence Council likewise states–

> expansion of nuclear power generation by 2025 to cover anywhere near the increasing demand would be virtually impossible. The infrastructure (human and physical), legal (permitting), and construction hurdles are just too big. Only at the end of our 15-20 year period are we likely to see a serious ramp up of nuclear technologies[66]

Faced with these divergent assessments, experts disagree on the ability of nuclear power to help meet CO_2 mitigation goals.

> Members of the Nuclear Power Joint Fact Finding (NJFF) reached no consensus about the likely rate of expansion for nuclear power in the world or in the United States over the next 50 years. Some group members thought it was unlikely that overall nuclear capacity would expand appreciably above its current levels and could decline; others thought that the nuclear industry could expand rapidly enough to fill a substantial portion of a carbon-stabilization "wedge" during the next 50 years.[67]

Consequently, congressional policy makers face a key uncertainty as to the achievable pace of nuclear power industry expansion, and therefore, the potential contribution nuclear generation may make to near-term CO_2 reduction.

ADVANCED COAL-FIRED POWER GENERATION

One set of technologies that may help to reduce CO_2 emissions from coal power plants is advanced coal-fired power generation. Advanced coal technologies include ultra-supercritical pulverized coal plants and integrated gasification combined cycle (IGCC) units, the latter of which convert coal into a synthetic gas prior to combustion. Compared to conventional coal-fired power plants, which typically operate at around 33% efficiency, advanced technologies under development or demonstration could improve coal plant efficiency to 46% or more.[68] Because they use coal more efficiently, advanced coal plants could yield proportionate reductions in CO_2 emissions per unit of electricity output compared to conventional coal plants. If carbon capture and sequestration technology (discussed in the next section) could be added to these plants their CO_2 emissions could be further reduced.

Some analysts and policy makers anticipate that advanced coal generation, either in the form of new coal-fired plants or upgrades to certain existing plants, will make a significant contribution to U.S. carbon reduction goals. They argue that, until carbon capture and storage technology is commercialized, advanced coal plants will be a cost-effective option for meeting future electricity demand growth while limiting CO_2 emissions growth compared to its current rate of growth. EPRI, for example, proposes CO_2 reductions from advanced coal plants on par with those from customer energy-efficiency programs through 2030.[69]

Uncertainty in Coal Plant Financing and Approval

While power plants employing advanced coal-fired generation technology face important questions about technological readiness and cost-effectiveness, the key uncertainty is whether they can be built—that is, whether the capital markets will finance them and whether regulators will permit them. Although they are more efficient than traditional coal plants, advanced technology coal plants still burn coal and—absent carbon capture technology—still release large volumes of CO_2 to the atmosphere. Once constructed, they may remain in service for 40 years or more. Hence they may not satisfy regulatory objectives for carbon control and may face financial risks stemming from future policies imposing costs on CO_2 emissions. Furthermore, industry arguments that these plants will be retrofitted with carbon capture technology when it becomes available are assuming the availability of that technology, which faces uncertainties as discussed in the following section.

Due to the potential carbon-related risks faced by new coal generation projects, major financial institutions are imposing greater requirements on developers seeking capital for new coal plant investments.[70] Regulatory agencies also have begun withholding regulatory approval from advanced coal project proposals. For example, in August 2007, the Minnesota Public Utilities Commission rejected a developer's proposal to construct a new IGCC power plant in the state as "not in the public interest."[71] The Sierra Club reports that at least 30 other proposed IGCC or supercritical coal generation projects have been cancelled across the country over the last several years due to financial and carbon emissions concerns.[72] In November, 2008, the Environmental Protection Agency's (EPA) Environmental Appeals Board ruled that an EPA region could not issue a permit for proposed coal-fired power plant without considering whether the "best available" CO_2 controls should be required for such a plant.[73] According to industry analysts, the EPA ruling would place a "freeze on the construction of as many as 100 new coal-fired power plants around the U.S."[74] The EPA Administrator has subsequently overruled the board's decision, apparently clearing the way for numerous coal plant permit applications to proceed, but raising new questions about future regulatory treatment of coal plant emissions under the next administration.[75]

CARBON CAPTURE AND SEQUESTRATION

Another approach to mitigating atmospheric carbon emissions is direct sequestration of carbon dioxide: capturing CO_2 at its source and storing it indefinitely to avoid its release to the atmosphere. Carbon capture and sequestration (CCS) is of great interest because potentially large amounts of CO_2 emitted from the industrial burning of fossil fuels in the United States could be suitable for sequestration. In theory, carbon capture technologies are seen as potentially removing 80%-95% of CO_2 emitted from an electric power plant or other industrial facility. Power plants are the most likely initial candidates for CCS because they are predominantly large, single-point sources, and they contribute approximately one-third of U.S. CO_2 emissions from fossil fuels.

Many analysts and policy makers have high hopes for CCS technology to help meet future CO_2 reduction goals in the U.S. electricity sector and worldwide. For example, one expert has testified before Congress that "[c]apturing and sequestering CO_2 emissions from coal-fired power plants and eventually all fossil combustion is a foundational technology component of any emissions reduction plan" seeking to stabilize atmospheric CO_2.[76] In the International Energy Agency's 2007 forecast CCS is "widely-deployed" under global CO_2 stabilization assumptions and accounts for 21% of avoided CO_2 emissions by 2030.[77] In EPRI's analysis, CCS makes the single greatest contribution to reduced CO_2 emissions among all measures by 2030.[78]

CCS Technology Uncertainty

Developing technology to capture CO_2 in an environmentally, economically, and operationally acceptable manner—especially from coal-fired power plants—has been an ongoing interest of the federal government for a decade. Nonetheless, the technology on the whole is still under development: no commercial device is currently available to capture carbon from coal plants. (Technology is currently available for capturing CO_2 from certain industrial processes, such as ammonia production, or natural gas processing, but the volumes of CO_2 emitted from coal-fired power plants dwarf any current industrial process in use today.) Various new CCS technologies have been tested successfully in laboratories, and some are being demonstrated in limited capacity field trials, but they are costly and face technical challenges to reaching commercial scale. Consequently, analysts disagree as to whether, when, and how CCS technology might be widely available in the United States at a cost competitive with commercially available technologies for generating electricity—particularly natural gas turbines.[79] Policies that impose costs on generators for CO_2 emissions, such as carbon taxes or cap-and-trade programs, presumably, would benefit CCS competitiveness, but they, too, are highly uncertain and constrained by other economic factors. As one study notes, "[t]he success of a cap-and-trade program in spurring widespread CCS deployment depends on a wide range of factors that cannot be controlled or even predicted in advance."[80]

A review of recent studies reveals a broad range of opinion as to how quickly CCS could be commercially deployed on large scale. An analysis from Imperial College, London, sees initial commercial deployment of CCS technology in the 2015-2020 period.[81] DOE officials reportedly have stated that the agency hopes to have carbon capture technology "in a deployable position within a decade."[82] A 2008 study by McKinsey & Company projected that early commercial CCS projects would be built "around 2020."[83]

The World Business Council for Sustainable Development believes CCS technology will enter commercial deployment by 2025.[84] The Intergovernmental Panel on Climate Change 2005 report on CCS technology states that "many integrated assessment analyses ... foresee the large-scale deployment of CCS systems within a few decades from the start of any significant regime for mitigating global warming."[85] A coalition including Greenpeace, Friends of the Earth, and other public interest groups asserts that "the best-case scenario is that [CCS] technology would be ready by 2030."[86] Referring to IEA's long term projections for carbon control, the OECD states that "there must be significant doubt as to whether or not it is feasible to achieve the assumed contribution from CCS by 2030."[87]

Given the range of expectations for CCS technology readiness, policy makers may not be able to establish to their satisfaction the contribution this technology may make to CO_2 emissions abatement. As one former DOE official reportedly has cautioned, "[p]eople's conception of how much [CCS] can deliver is probably wildly overrated. So I'm hopeful that carbon capture and storage could be part of the solution, but ... one just has to put a big question mark on it."[88] Echoing this view, Nobel laureate Steven Chu, Director of the Lawrence Berkeley National Laboratory and nominee for Secretary of Energy, reportedly has remarked, "it's not guaranteed we have a solution for coal."[89] Accordingly, the key uncertainty for policy makers regarding CCS is if and when the technology might be available for widespread deployment.

PLUG-IN ELECTRIC HYBRID VEHICLES

A recent development in advanced vehicle technologies is the potential introduction in the next few years of plug-in hybrid electric vehicles (PHEVs). PHEVs, like commercially available hybrid electric vehicles (HEVs), would combine an electric motor and battery pack with an internal combustion engine to improve overall fuel efficiency. PHEVs would use a much higher-capacity battery pack than a typical HEV, however, and would be able charge the vehicle on grid power rather than solely by the combustion engine during operation. With their larger batteries, PHEVs could achieve an all-electric range of 20 to 40 miles (the average commuting distance). By using grid electricity rather than gasoline in this way, PHEVs could lower overall fuel-cycle pollutant emissions—including CO_2 emissions—from vehicles.[90] Accordingly, many analysts and policy makers believe PHEV's offer substantial opportunities to reduce the nation's overall CO_2 emissions. A 2007 joint EPRI-Natural Resources Defense Council (NRDC) study, for example, concluded that, through the widespread adoption of PHEV technology, "cumulative [CO_2] savings from 2010 to 2050 can be large."[91] The Obama-Biden presidential campaign pledged to "[p]ut 1 million plug-in hybrid cars ... on the road by 2015."[92]

While PHEVs are an innovative technology, there are important questions about the impact PHEVs may have on CO_2 emissions over the next 20 ot 30 years. One key uncertainty is whether such vehicles would be purchased in sufficient numbers (approximately 50% of new car sales by 2025, according to the EPRI/NRDC study)[93] to have a significant carbon impact despite their high cost. PHEVs are projected to retail for over $40,000 per vehicle in the near-term compared to approximately $28,000 for an HEV and $23,000 for a conventional vehicle.[94] As a point of reference, conventional HEVs accounted for 2.4% of new vehicle sales in the United States through November 2008.[95] Of greater uncertainty, perhaps, are the projected carbon emissions of power plants operating to supply PHEV electricity. As a general rule, PHEVs only reduce net carbon emissions if the power plants supplying them produce relatively little carbon per kWh. But some studies show that, if the U.S. generation portfolio does not significantly reduce its overall carbon intensity, widespread adoption of PHEVs through 2030 may have only a small effect on, and might actually increase, net CO_2 emissions.[96] Thus, the carbon abatement potential of PHEVs is largely dependent upon the concurrent implementation of renewables, nuclear power, and CCS—each of which face great uncertainties of their own as discussed above.

DISTRIBUTED ENERGY RESOURCES

Distributed energy resources are small-scale power generation technologies located near homes or businesses to provide an alternative to, or an enhancement of, conventional grid power. Distributed resources include technologies such as rooftop photovoltaics, natural gas-fired microturbines, wind turbines, and fuel cells. The category also includes combined heat and power (CHP) systems, which make productive use of "waste" heat from electricity generation, thereby increasing the total useful energy extracted from electric generation fuels.[97] Distributed energy resources (DER) offer potential benefits to customers in terms of energy costs, power reliability, and power quality. DER technologies also can help mitigate CO_2 emissions because some use renewable energy sources, or, as in the case of CHP systems, they make more efficient use of fossil fuels than the utility power generation they displace.[98]

Among DER technologies, CHP is the most widespread, accounting for nearly nine percent of U.S. electric generating capacity in 2007.[99] CHP is also viewed as having the greatest near-term potential to reduce CO_2 emissions. A 2008 report from Oak Ridge National Laboratory concludes that, by achieving 20 percent generation capacity from CHP by 2030 (an "aggressive target"), the United States could avoid 60 percent of the projected increases in CO_2 emissions during that time.[100] A 2007 study by McKinsey & Company further concludes that much of this added capacity could be installed at negative marginal cost.[101] Nonetheless, CHP, along with other DER technologies, has not been implemented to its potential due to technical and utility infrastructure barriers.

> CHP, or cogeneration, has been around in one form or another for more than 100 years; it is proven, not speculative. Despite this proven track record, CHP remains underutilized and is one of the most compelling sources of energy efficiency that could, with even modest investments, move the Nation strongly toward ... a cleaner environment.[102]

Other DER technologies, such as photovoltaics and fuel cells, likewise face barriers limiting their implementation. As a United Kingdom government study stated,

> The complexity and novelty of some of the technologies, together with their need to be integrated into the built environment, often by players new to the energy business, means there is a significant gap between potential and delivery. Moreover, many of the technologies are not yet cost-competitive at their current state of development and with current fuel and carbon prices.[103]

Analysts also question whether policies imposing costs on CO_2 emissions would substantially increase DER adoption due to the existing alignment of carbon reduction and cost reduction objectives, even without carbon costs, and the fundamental economics of renewables.[104] Thus, DER faces a key uncertainty similar to that faced by energy efficiency. The key question is whether new programs and the imposition of carbon costs would enable the electricity sector to capture substantially more DER potential than it does today.

POLICY ISSUES FOR CONGRESS

Policy research and technical studies show that substantially reducing CO_2 emissions in the U.S. electricity sector over the next few decades likely requires successful deployment of every major carbon mitigation measure at the nation's disposal. However, it is also clear that significant uncertainties cast doubt on the potential of individual measures to achieve their hoped-for carbon impact. For the measures discussed in this report, the key uncertainties can be summarized as follows:

- **Energy efficiency**—Can the United States overcome socioeconomic barriers to achieve four times more potential savings than ever before?
- **Renewable energy**—Will there be enough transmission for wind power? Is there enough land to grow the needed biomass?
- **Nuclear power**—Could the United States build new plants fast enough to matter?
- **Advanced coal power**—Will banks fund them and regulators approve them?
- **Carbon capture and sequestration**—Will the technology be commercially deployable in 10 years, 25 years, or never?
- **Plug-in hybrid electric vehicles**—How much "low carbon" electricity would be available to charge their batteries?
- **Distributed energy resources**—Would carbon costs change distributed energy economics enough to spur deployment?

Policy makers and interest groups recognize these uncertainties, and have put forth numerous proposals to address them. It is beyond the scope of this report to examine each of these proposals, but they include the broadest range of policy instruments at government's disposal: higher efficiency standards, new regulatory authorities, tax incentives, direct subsidies, research and development (R&D) grants, environmental rules, public information campaigns, and a host of other policy instruments. Specific examples include calls for more federally-funded CCS demonstration projects and proposals for federal preemption of state siting authority to promote new transmission development.[105] While they run the gamut, it remains to be seen which proposals may be pursued by Congress and what effects they would have. Consequently, the overall success of a multi-measure CO_2 mitigation scheme such as that proposed by EPRI in **Figure 1**, and its economic underpinnings, is inherently unpredictable.

> The cost of building and operating coal plants with and without CCS systems, the cost of natural gas, nuclear power and renewable sources of power, the cost of emissions offsets from outside the utility sector, and ultimately the market price of CO_2 itself are all variables that will dictate the decisions of future power plant developers. These variables are all highly uncertain from today's perspective and may create a set of economic drivers dramatically different from those anticipated by policymakers.[106]

Possible Outcomes for Carbon Control

Identifying key uncertainties for carbon abatement measures is not the same as predicting their ultimate success or failure. This report does not independently assess the likelihood of particular measures meeting any specific CO_2 reduction target. The whole point of this report's focus on uncertainty is that CO_2 outcomes are not known, and expert opinions vary as to what the future will be. It is, therefore, entirely possible that a portfolio of CO_2 measures such as those in **Figure 1**, implemented under the optimum subset of policies currently under debate, could achieve the types of carbon reductions projected. Indeed, EPRI has characterized its CO_2 scenario as "very aggressive, but potentially feasible."[107] Other CO_2 abatement analyses have also been developed in good faith with a similar belief in their practicality. Successfully reducing U.S. carbon emissions to 1990 levels, say, by 2030 would validate the overall policy approach as well as the specific measures comprising it. Such an outcome would not be conclusive, as deeper CO_2 cuts would arguably need to follow, but the intervening years would probably provide greater clarity on the best policy options to 2050 and beyond.

Underperformance of Individual CO Measures

Although successful implementation of all carbon abatement measures is the goal, it is also entirely possible that, under a multi-measure strategy like EPRI's, one or more measures would fall short of meeting expectations for CO_2 reduction. Nuclear industry expansion, for instance, could easily fail to materialize, or the transmission grid could expand too slowly for sustained wind power development. In such a case legislators might need to revisit both the enabling policies for the underperforming measures and for the carbon strategy as a whole.

One obvious solution could be to rely on more successful measures to compensate for the underperforming ones. As the OECD study posits, "if CCS and/or end-use efficiency fail to achieve the required targets, it follows that other technologies, including nuclear power, will need to make bigger contributions to fill the gap."[108] But based on a review of the relevant research cited in this report, such an approach may simply not be realistic. Since the CO_2 targets of the individual measures are already viewed by some analysts as "ambitious," "aggressive," or " a major challenge," and there are questions as to whether the initial targets can be achieved, increasing those targets may stretch credibility and sharply increase the uncertainty of the overall CO_2 mitigation effort. Alternatively, legislators could revisit measure-specific policies to see if more intervention could improve their particular prospects. Larger tax credits for renewable energy, for example, could potentially improve the competitiveness of biomass generation, thereby encouraging biomass investment. Such an approach might still encounter difficulty making up for early underperformance of a measure in the out years, as CO_2 emissions are cumulative, but any improvement would arguably be helpful. A third option would be to increase the level of any future carbon costs with the expectation that an increase would benefit carbon-mitigating technologies across the board. Such an action, however, could have broad implications for energy prices, reliability, and availability.

Failure of the CO Mitigation Portfolio

Notwithstanding the best efforts of federal policy makers, it is possible that, given the uncertainties they face, that few if any of the major measures proposed to moderate U.S. carbon emissions would achieve their anticipated impacts in a 20-year time frame. In its 2008 report, the National Intelligence Council suggests just such an outcome:

> [A]ll current technologies are inadequate for replacing the traditional energy architecture on the scale needed, and new energy technologies probably will not be commercially viable and widespread by 2025.... Even with a favorable policy and funding environment for biofuels, clean coal, or hydrogen, the transition to new fuels will be slow.[109]

Under such a scenario, legislators may face different policy alternatives, each with distinct but potentially significant implications. Congress might increase future carbon emissions costs even higher than under the previous scenario—to a level that virtually guarantees the targeted CO_2 reductions—although the effects of very high carbon prices on electricity costs might be detrimental to specific industries (e.g., coal) or to the economy as a whole. Alternatively, Congress might refocus its efforts on a single measure (e.g.,

CCS or nuclear power) restructuring its policies to expand the implementation of that specific measure far beyond its original targets. Congress could also reset its national CO_2 targets, deferring reduction goals until they are more in line with the maturation and implementation of the key CO_2 reduction technologies. In this case, legislators would signal acceptance that near-term targets could not be met and hope for greater success later in the century. This alternative risks unacceptable implications, however, and might violate future international emissions treaties. Finally, Congress could abandon its focus on CO_2 reduction altogether, instead directing resources at mitigating the effects of global warming and adapting to a hotter climate.

CONCLUSION

Reducing U.S. emissions of manmade CO_2 is a priority of both the President and leaders in Congress. Comprehensive policies have been proposed to achieve these reductions. Most envision aggressive implementation of a portfolio of major carbon reduction measures, with the goal of reducing U.S. CO_2 emissions to 1990 levels by 2020 or 2030. Numerous studies support the potential of specific measures to lower CO_2 emissions, but also identify key implementation uncertainties which may impact their overall viability. Congress is considering policies to address these uncertainties, but which policies may be implemented and how effective they may be cannot be known at this time.

As the nation's CO_2 mitigation policies continue to develop, the inherent uncertainty associated with specific carbon measures may be a critical concern. Commitments, either domestic or international, to specific carbon emissions targets over time, or to a specific schedule of carbon costs (whatever form they may take) may be greatly affected by the success of the underlying measures relied upon to achieve them. The reverse is also true; a schedule of carbon costs may also influence the success of CO_2 abatement measures. Therefore, policy makers may benefit from a complete and integrated understanding of measure-specific uncertainties and the range of carbon outcomes they imply. As one study has concluded, "explicitly including uncertainty in both technical change and in climate damages is important for understanding the relationship between technical change, climate change, and policy.... [O]ptimal policy is different ... when uncertainty is taken into account.[110]

As Congress considers implementing CO_2 policies, keeping a close eye on the technology and market developments associated with every key measure may be an oversight priority. Success or failure of any particular measure may be apparent early on in its implementation, affording an opportunity through quick action to make policy adjustments to improve a measure's chances for success, or to abandon it in favor of more promising options. Perhaps more importantly, given the complexity and scale of the carbon control problem, Congress may also find it useful to expect the unexpected in the electricity sector. Apart from the measures discussed in this report, new technologies, consumer behavior, or infrastructure developments (e.g., a rush to natural gas) may emerge rapidly and unexpectedly to change fundamental aspects of the nation's carbon emissions trajectory. One way or another, electricity supply will balance with demand—but perhaps in unanticipated ways. The recent volatility in global oil prices is a relevant example of unexpected structural changes in energy markets. Balancing responses to energy market volatility and unexpected structural changes against the need for a predictability in R&D and private capital investment may be essential to maintaining the nation on course to meaningful atmospheric CO_2 reduction.

End Notes

[1] World Resources Institute, "Comparison of Legislative Climate Change Targets," (Washington, DC: June 18, 2008): 3.

[2] President-elect Barack Obama, Remarks before the Governor's Global Climate Summit, Beverly Hills, California (November 18, 2008).

[3] For another prominent analysis, see also: S. Pacala and R. Socolow, "Stabilization Wedges: Solving the Climate Problem for the Next 50 Years with Current Technologies," *Science*, Vol. 305, No. 5686 (August 13, 2004): 968 - 972.

[4] See, for example: Seung-Rae Kim, Klaus Keller, and David F. Bradford, "Optimal Technological Portfolios for Climate-Change Policy Under Uncertainty: A Computable General Equilibrium Approach," Presented at the 10th International Society of Computational Economics Conference on Computing in Economics and Finance (Amsterdam: July 8-10, 2004).

[5] Strictly speaking, "conservation" means "avoiding waste," but the term is typically used interchangeably with "efficiency" in the energy policy context, as it is in this report. "Efficiency" and "conservation" contrast with "curtailment" or "load management" which decrease output (e.g., turning down the thermostat) or services (e.g., driving less) to decrease energy use at specific times. Curtailment is often employed as an emergency measure.

[6] Amory B. Lovins, "Energy Strategy: The Road Not Taken?" *Foreign Affairs*, Vol. 55 No. 1 (October 1976).

[7] Now the National Renewable Energy Laboratory.

[8] Solar Energy Research Institute (SERI), *A New Prosperity: Building a Sustainable Energy Future*, Brick House Publishing (Andover, MA: 1981): 1.

[9] Ibid: 2.

[10] Interlaboratory Working Group on Energy-Efficient and Low-Carbon Technologies, *Scenarios of U.S. Carbon Reductions: Potential Impacts of Energy-Efficient and Low-Carbon Technologies by 2010 and Beyond*, (1997) :3.3-3.4. http://enduse.lbl.gov/projects/5lab.html The five laboratories are Oak Ridge National Laboratory, Lawrence Berkeley National Laboratory, Pacific Northwest Laboratory, Argonne National Laboratory, and the National Renewable Energy Laboratory.

[11] Steven Nadel, Anna Shipley and R. Neal Elliott, "The Technical, Economic and Achievable Potential for Energy-Efficiency in the U.S. – A Meta-Analysis of Recent Studies," *Proceedings of the 2004 ACEEE Summer Study on Energy Efficiency in Buildings*, American Council for an Energy-Efficient Economy (Washington, DC: 2004).

[12] U.S. Department of State, *U.S. Climate Action Report—2006* (July 2007): 40.

[13] Commonly referred to as "demand-side management" or "DSM" programs.

[14] Kenneth Gillingham, Richard G. Newell, and Karen Palmer, *Retrospective Examination of Demand-Side Energy Efficiency Policies*, Resources for the Future, RFF DP 04-19 REV (June 2004; revised September 2004): 63-64.

[15] Charles A. Goldman, Joseph H. Eto, and Galen L. Barbose, *California Customer Load Reductions during the Electricity Crisis: Did they Help to Keep the Lights On?*, Lawrence Berkeley National Laboratory, LBNL-49733 (May 2002): iii, 20.

[16] Michael Howard, Senior Vice President, "Electric Power Research Institute, Energy Efficiency: How Much Can We Count On?" Presented at the Edison Foundation Conference, Keeping the Lights On: Our National Challenge, (April 21, 2008): 14. http://www.edisonfoundation.net/events/2008-04-21/EPRIPresentation.pdf

[17] J.G. Koomey, C.A. Webber, C.S. Atkinson, and A. Nicholls, "Addressing Energy-Related Challenges for the U.S. Buildings Sector: Results from the Clean Energy Futures Study," Energy Policy, Vol. 29, No. 14 (November 2001): 1211.

[18] Howard Geller and Sophie Attali, "The Experience with Energy Efficiency Policies and Programmes in IEA Countries: Learning from the Critics," International Energy Agency, IEA Information Paper (August 2005): 23.

[19] William H. Golove and Joseph H. Eto, Market Barriers to Energy Efficiency: A Critical Reappraisal of the Rationale for Public Policies to Promote Energy Efficiency, LBL-38059, UC-1322, Lawrence Berkeley National Laboratory (March 1996): xii-xiii.

[20] For further analysis of conservation proposals, see CRS Report RL33831, *Energy Efficiency and Renewable Energy Legislation in the 110th Congress*, by Fred Sissine, Lynn J. Cunningham, and Mark Gurevitz.

[21] This report excludes hydroelectric power generation from the renewables category because of environmental constraints and associated limits on potential new hydro capacity.

[22] T. Boone Pickens, *The Plan*, Internet page (August 7, 2008). http://www.pickensplan.com/theplan

[23] Obama-Biden, *Barack Obama and Joe Biden: New Energy for America*, Fact sheet (August 3, 2008). http://www.barackobama.com/pdf/factsheet_energy_speech_080308.pdf; A 2004 Illinois state senate bill cosponsored by Senator Barack Obama (SB2321, 93rd General Assembly) excluded new construction hydroelectric power from its proposed state RPS.

[24] See, for example: American Council On Renewable Energy, The Outlook on Renewable Energy in America Volume II: Joint Summary Report (Washington, DC: March 2007).

[25] Barbara Tyran, Electric Power Research Institute, (May 15, 2008): 7.

[26] U.S. Energy Information Administration, *Annual Energy Outlook 2008* (June 2008): 70.

[27] For a more comprehensive analysis of wind power, see CRS Report RL34546, *Wind Power in the United States: Technology, Economic, and Policy Issues*, by Jeffrey Logan and Stan Mark Kaplan.

[28] Federal tax credits are provided for under the Energy Policy Act of 1992 (P.L. 102-486 § 1914) and subsequent statutes.

[29] U.S. Energy Information Administration, *Electric Power Monthly*, DOE/EIA-0226 (2008/10) (October 28, 2008): Tables 1.1 and 1.1a. This figure is year to date through July.

[30] D. L. Elliott, L. L. Wendell, G.L. Gower, An Assessment of the Available Windy Land Area and Wind Energy Potential in the Contiguous United States 1991, Pacific Northwest Laboratory, PNL-7789 (August, 1991): iii.

[31] Walter Short and Nate Blair, The Long-Term Potential of Wind Power in the U.S., *Solar Today* (November/December 2003): 29.

[32] U.S. Department of Energy, 20% Wind Energy by 203 0: Increasing Wind Energy's Contribution to U.S. Electricity Supply, DOE/GO-102008-2567 (July 2008): 1.

[33] U.S. Department of Energy (July 2008): 14.

[34] Ibid.

[35] National Grid Corp., *Transmission and Wind Energy: Capturing the Prevailing Winds for the Benefit of Customers* (Westborough, MA: September 2006): 9. http://www.nationalgridus.com/non_html/c3-3_NG_wind_policy.pdf

[36] U.S. Department of Energy (July 2008): 95. This compares to 200,000 miles of existing transmission lines operating at 230 kV and above.

[37] AEP, "Interstate Transmission Vision for Wind Integration," Electricity Today (September 2007): 30-37.

[38] Shalini P. Vajjhala, "Siting Difficulty and Transmission Investment,"IAEE Energy Forum (International Association for Energy Economics: 2nd Quarter 2008): 5-7; North American Electric Reliability Council, 2006 Long-Term Reliability Assessment (October 2006): 22-23.

[39] U.S. Dept. of Energy, *Wind Energy Multiyear Program Plan For 2007–2012*, DOE/GO-102007-2451 (August 2007): 8.

[40] Matthew L. Wald, "Wind Energy Bumps into Power Grid's Limits," *New York Times* (August 27, 2008).

[41] Additional CO_2 is emitted by combustion of fossil fuels used to produce and process fuel crops, but those emissions are small compared to power plant combustion emissions.

[42] U.S. Energy Information Administration, *Electric Power Monthly*, DOE/EIA-0226 (2008/10) (October 28, 2008): Tables 1.1 and 1.1a. This value is year-to-date through July 2008.

[43] U.S. Energy Information Administration, *Biomass for Electricity Generation* (2002). http://www.eia.doe.gov/oiaf/ analysispaper/biomass/ The study excludes biomass and coal co-firing.

[44] U.S. Department of Agriculture, Biomass as a Feedstock for a Bioenergy and Bioproducts Industry: The Technical Feasibility of a Billion-Ton Annual Supply, (April 2005): 34.

[45] Michael Toman, James Griffin, and Robert J. Lempert, Impacts on U.S. Energy Expenditures and Greenhouse-Gas Emissions of Increasing Renewable-Energy Use, RAND Corp. (2008): 38.

[46] Michael Toman, et al. (2008): xiii.

[47] For further analysis of renewable fuels issues, see CRS Report RL34265, *Selected Issues Related to an Expansion of the Renewable Fuel Standard (RFS)*, by Brent D. Yacobucci and Tom Capehart.

[48] U.S. Environmental Protection Agency, *Greenhouse Gas Mitigation Potential in U.S. Forestry and Agriculture*, EPA 430-R-05-006 (November 2005): 4-12.

[49] John Reilly and Sergey Paltsev, "Biomass Energy and Competition for Land," MIT Joint Program on the Science and Policy of Global Change, Report No. 145 (April 2007):16.

[50] U.S. Nuclear Regulatory Commission, *Information Digest 2008-20098*, NUREG-1350, Vol. 20 (August 2008): 32.

[51] U.S. Energy Information Administration, *Status of Potential New Commercial Nuclear Reactors in the United States* (October 9, 2008): Table 1. http://www.eia.doe.gov/cneaf/ nuclear/page/nuc_reactors/com_reactors.pdf

[52] Processes related to nuclear generation, such as the mining, processing, and transportation of nuclear fuel, do generate CO_2 emissions. The levels of these emissions are the subject of debate. See Kurt Kleiner, "Nuclear Energy: Assessing the Emissions," *Nature Reports: Climate Change* (September 24, 2008). http://www.nature.com/climate/2008/0810/pdf/ climate.2008.99.pdf

[53] President-elect Barack Obama (November 18, 2008).

[54] Intergovernmental Panel on Climate Change (IPCC), Climate Change 2007: Mitigation, Contribution of Working Group III to the Fourth Assessment Report of the Intergovernmental Panel on Climate Change (Cambridge University Press: 2007): 253.

[55] Barbara Tyran, Electric Power Research Institute, (May 15, 2008): 7.

[56] Massachusetts Institute of Technology (MIT), *The Future of Nuclear Power* (2003): 3.

[57] Natural Resources Defense Council, *Nuclear Facts*, (February 2007): 1. http://www.nrdc.org/ nuclear/plants/ plants.pdf

[58] IPCC (2007): 253; MIT (2003): 21-24.

[59] U.S. Department of Energy, A Roadmap to Deploy New Nuclear Power Plants in the United States by 2010: Volume 1 Summary Report (October 31, 2001): 7.

[60] Organisation for Economic Co-operation and Development (OECD), *Nuclear Energy Agency, Nuclear Energy Outlook 2008*, NEA No. 6348(2008): 316. GWe=Gigawatt-electric.

[61] Nuclear Energy Advisory Committee, Nuclear Energy: Policies and Technology for the 21st Century (November 2008): 8.

[62] Compiled from the Energy Information Administration (EIA) Reactor Status List, available at http://www.eia.doe.gov/cneaf/nuclear/page/nuc_reactors/reactsum.html.

[63] Mycle Schneider, "2008 World Nuclear Industry Status Report: Global Nuclear Power," *Bulletin of the Atomic Scientists*, Web edition (September 16, 2008). http://www.thebulletin.org/web-edition/reports/2008-world-nuclearindustry-status-report/2008-world-nuclear-industry-status-rep

[64] Sharon Squassoni, *Nuclear Renaissance: Is It Coming? Should It?* Carnegie Endowment for International Peace (October 2008): 2.

[65] U.S. Secretary of Energy Samuel Bodman, Remarks before the Nuclear Energy Summit: *Renewing America's Nuclear Power Partnership For Energy Security and Economic Growth*, Convened at the U.S. Department of Commerce (October 8, 2008).

[66] National Intelligence Council, *Global Trends 2025: A Transformed World*, NIC 2008-003 (November 2008): 45.

[67] Keystone Center, *Nuclear Power Joint Fact-Finding* (June 2007): 10.

[68] Massachusetts Institute of Technology (MIT), *The Future of Coal* (2007): 115, 124.

[69] Barbara Tyran, Electric Power Research Institute, (May 15, 2008): 7.

[70] Citi, JPMorgan Chase and Morgan Stanley, "Leading Wall Street Banks Establish The Carbon Principles," Joint press release (New York: February 4, 2008).

[71] Minnesota Public Utilities Commission, Order Resolving Procedural Issues, Disapproving Power Purchase Agreement, Requiring Further Negotiations, and Resolving to Explore the Potential for a Statewide Market for Project Power under Minn. Stat. § 216b.1694, Subd. 5, Docket No. E-6472/M-05-1993 (August 30, 2007):15.

[72] Sierra Club, "Stopping the Coal Rush," Internet database (December 16, 2009). http://www.sierraclub.org/ environmentallaw/coal/plantlist.asp

[73] U.S. Environmental Protection Agency, Environmental Appeals Board, *In re: Deseret Power Electric Cooperative PSD Permit No. PSD-OU-0002-04.00*, PSD Appeal No. 07-03 Order Denying Review in Part and Remanding in Part (Decided November 13, 2008).

[74] Kate Sheppard, "Is That a Bonanza in Your Docket?," *Gristmill*, Internet blog (November 14, 2008). http://gristmill.grist.org/story/2008/11/13/165551/28

[75] Stephen L. Johnson, Administrator, U.S. Environmental Protection Agency, EPA's Interpretation of Regulations that Determine Pollutants Covered By Federal Prevention of Significant Deterioration (PSD) Permit Program, Memorandum (December 18, 2008).

[76] Raymond J. Kopp, Director of the Climate Policy Program, Resources for the Future, Testimony before the Senate Energy and Natural Resources Committee (June 25, 2008).

[77] International Energy Agency (IEA), *World Energy Outlook 2007* (2007): 208.

[78] Barbara Tyran, Electric Power Research Institute, (May 15, 2008): 7.

[79] For discussion of specific carbon capture technologies and analysis of the potential costs implications of carbon capture relative to other generation technologies, see CRS Report RL34621, *Capturing CO_2 from Coal-Fired Power Plants: Challenges for a Comprehensive Strategy*, by Larry Parker, Peter Folger, and Deborah D. Stine.

[80] Ken Berlin and Robert M. Sussman, *Global Warming and The Future of Coal*, Center for American Progress (May 2007): 34.

[81] J. Gibbins and H. Chalmers, "Preparing for Global Rollout: a 'Developed Country First' Demonstration Programme for Rapid CCS Deployment," *Energy Policy*, Vol. 36, No. 2 (2008): 501-507.

[82] Acting Deputy Energy Secretary Jeffrey Kupfer quoted in "Bush DOE Transition Head Sees Commercial Carbon Capture In 10 Years," *EnergyWashington* (December 2, 2008).

[83] McKinsey & Company, Inc., *Carbon Capture & Storage: Assessing the Economics* (September 22, 2008): 18.

[84] World Business Council for Sustainable Development, *Pathways to 2050: Energy and Climate Change* (December 7, 2005): 5.

[85] Intergovernmental Panel on Climate Change (IPCC), Special Report: *Carbon Dioxide Capture and Storage*, 2005 (2005): 341

[86] Greenpeace, "Tell Congress To Keep Carbon Capture and Storage Out Of Energy Legislation," Internet page (Washington, DC: May 6, 2008) http://members.greenpeace.org/ action/start.php?action_id=189; "Public Interest Groups Oppose Carbon Capture Scam," *It's Getting Hot in Here*, Internet blog (May 6, 2008). http://itsgettinghotinhere.org/ 2008/05/06/public-interest-groups-oppose-carbon-capture-scam/

[87] OECD (2008): 133.

[88] Joseph Romm, Center for American Progress, as quoted in "Coal's Carbon Capture in Question," *Living on Earth*, Public Radio International, audio transcript (February 8, 2008).

[89] Keith Johnson, "Steven Chu: 'Coal is My Worst Nightmare'," *Wall Street Journal, Environmental Capital*, Internet blog (December 11, 2008). http://blogs.wsj.com/ environmentalcapital/2008/12/11

[90] While these emissions reductions, technically, affect the transportation sector, they are linked to emissions in the electricity sector and so are often assessed in both contexts.

[91] Electric Power Research Institute and Natural Resources Defense Council, Environmental Assessment of Plug-In Hybrid Electric Vehicles, Volume 1: Nationwide Greenhouse Gas Emissions (July 2007): 5-10.

[92] Obama-Biden, *Barack Obama and Joe Biden: New Energy for America*, Fact sheet (August 3, 2008). http://www.barackobama.com/pdf/factsheet_energy_speech_080308.pdf

[93] Electric Power Research Institute and Natural Resources Defense Council (July 2007): 4-8. This value is for the "Medium PHEV fleet penetration case."

[94] Andrew Simpson, *Cost-Benefit Analysis of Plug-In Hybrid Electric Vehicle Technology*, National Renewable Energy Laboratory, NREL/CP-540-40485 (November 2006): 10.

[95] Green Car Congress, *US Sales of Hybrids Down 50% in November*, Web page (December 9, 2008). http://www.greencarcongress.com/2008/12/us-sales-of-hyb.html#more

[96] See, for example: Stanton W. Hadley and Alexandra Tsvetkova, *Potential Impacts of Plug-in Hybrid Electric Vehicles on Regional Power Generation*, Oak Ridge National Laboratory, ORNL/TM-2007/150 (January 2008): 68; Constantine Samaras and Kyle Meisterling, "Life Cycle Assessment of Greenhouse Gas Emissions from Plug-in Hybrid Vehicles: Implications for Policy," *Environmental Science and Technology*, Vol. 42, No. 9. (2008): 3170-3176.

[97] CHP systems are also commonly referred to as "cogeneration" systems.

98 S.W. Hadley, J.W. Van Dyke, W.P. Poore, III, and T.K. Stovall, *Quantitative Assessment of Distributed Energy Resource Benefits*, Oak Ridge National Laboratory, ORNL/TM-2003/20 (May 2003): 23.

99 Oak Ridge National Laboratory, Combined Heat and Power: Effective Energy Solutions for a Sustainable Future, ORNL/TM-2008/224 (December 1, 2008): 4.

100 Oak Ridge National Laboratory, (December 1, 2008): 21.

101 McKinsey & Company, Inc., Reducing U.S. Greenhouse Gas Emissions: How Much at What Cost?, (December, 2007).

102 Oak Ridge National Laboratory, *Combined Heat and Power: Effective Energy Solutions for a Sustainable Future*, ORNL/TM-2008/224 (December 1, 2008): 3.

103 United Kingdom, Department for Business Enterprise and Regulatory Reform, *UK Renewable Energy Strategy: Consultation Document* (2007): 135. http://renewableconsultation.berr. gov.uk/consultation/chapter-5/executive-summary/

104 Ryan Firestone and Chris Marnay, "Distributed Energy Resources for Carbon Emissions Mitigation," Conference paper for The European Council for an Energy Efficient Economy 2007 Summer Study, (La Colle sur Loup, France: June 4-9, 2007): 5.

105 Union of Concerned Scientists, *Coal Power in a Warming World* (October 2008): 3-4; Testimony of Joseph Kelliher, Chairman of the Federal Energy Regulatory Commission, before the Senate Energy and Natural Resources Committee (July 31, 2008). http://www.ferc.gov/EventCalendar/Files/20080731102123-Chairmantestimony.pdf

106 Ken Berlin and Robert M. Sussman (May 2007): 34.

107 Barbara Tyran, Electric Power Research Institute, (May 15, 2008): 7.

108 OECD (2008): 133.

109 National Intelligence Council, *Global Trends 2025: A Transformed World*, NIC 2008-003 (November 2008). http://www.dni.gov/nic/PDF_2025/2025_Global_Trends_ Final_Report. pdf

110 Erin Baker, Leon Clarke, Jeffrey Keisler, and Ekudayo Shittu, "Uncertainty, Technical Change and Policy Models," University of Massachusetts, Boston, College of Management, UMBCMWP 1028 (July 2007): 3-4.

In: Emissions Control in Electricity Generation ISBN: 978-1-62257-830-6
Editors: J. L. Walker, A. Hernandez

Chapter 2

AVAILABLE AND EMERGING TECHNOLOGIES FOR REDUCING GREENHOUSE GAS EMISSIONS FROM COAL-FIRED ELECTRIC GENERATING UNITS*

United States Environmental Protection Agency

ACRONYMS AND ABBREVIATIONS

APFBC	Advanced pressurized fluidized bed combustion
ASTM	American Society for Testing and Materials
ASME	American Society of Mechanical Engineers
ASU	Air separation unit
BACT	Best Available Control Technology
Btu	British thermal unit
CAA	Clean Air Act
CCS	Carbon capture and storage
CEMS	Continuous emission monitoring system
CFB	Circulating fluidized bed
CH4	Methane
CO	Carbon monoxide

* This is an edited, reformatted and augmented version of the United States Environmental Protection Agency publication, dated October 2010.

CO_2	Carbon dioxide
EGU	Electric generating unit
EPA	U.S. Environmental Protection Agency
FBC	Fluidized bed combustion
EPRI	Electric Power Research Institute
FGD	Flue gas desulfurization
GHG	Greenhouse gas
H2O	Water
HRSG	Heat recovery steam generator
HHV	Higher heating value
IGCC	Integrated gasification combined cycle
IEA	International Energy Agency
kJ	Kilojoule
kW	Kilowatt
kWh	Kilowatt-hour
LCOE	Levelized cost of electricity
Mg	Megagram
MMBtu/hr	Million Btu per hour
MPa	Megapascal
MW	Megawatt
MWe	Megawatt electrical
MWh	Megawatt-hour
MSW	Municipal solid waste
N2O	Nitrous oxide
NETL	National Energy Technology Laboratory
NOX	Nitrogen oxides
O&M	Operation and maintenance
PC	Pulverized coal
PFBC	Pressurized fluidized bed combustion
PM	Particulate matter
PRB	Power River Basin
scfm	Standard cubic feet per minute
SO_2	Sulfur dioxide
SO_3	Sulfur trioxide
SNCR	Selective noncatalytic reduction
ton/day	tons per day
ton/yr	tons per year
U.S. DOE	U.S. Department of Energy
U.S. EIA	U.S. Energy Information Administration

1. INTRODUCTION

This document is one of several white papers that summarize readily available information on control techniques and measures to mitigate greenhouse gas (GHG) emissions from specific industrial sectors. These white papers are solely intended to provide basic information on GHG control technologies and reduction measures in order to assist States and local air pollution control agencies, tribal authorities, and regulated entities in implementing technologies or measures to reduce GHGs under the Clean Air Act, particularly in permitting under the prevention of significant deterioration (PSD) program and the assessment of best available control technology (BACT). These white papers do not set policy, standards or otherwise establish any binding requirements; such requirements are contained in the applicable EPA regulations and approved state implementation plans.

This document provides information on control techniques and measures that are available to mitigate GHG emissions from the coal-fired electric generating sector at this time. The primary GHG emitted by the coal-fired electric generation industry is carbon dioxide (CO_2), and the control technologies and measures presented in this document focus on this pollutant. While a large number of available technologies are discussed here, this paper does not necessarily represent all potentially available technologies or measures that that may be considered for any given source for the purposes of reducing its GHG emissions. For example, controls that are applied to other industrial source categories with exhaust streams similar to the cement manufacturing sector may be available through "technology transfer" or new technologies may be developed for use in this sector.

The information presented in this document does not represent U.S. EPA endorsement of any particular control strategy. As such, it should not be construed as EPA approval of a particular control technology or measure, or of the emissions reductions that could be achieved by a particular unit or source under review.

1.1. Electric Power Generation Using Coal

Electricity is generated at most electric power plants by using mechanical energy to rotate the shaft of electromechanical generators. The mechanical energy needed to rotate the generator shaft can be produced from the conversion of chemical energy by burning fuels or from nuclear fission; from the conversion of kinetic energy from flowing water, wind, or tides; or from the conversion of thermal energy from geothermal wells or concentrated solar energy. Electricity also can be produced directly from sunlight using photovoltaic cells or by using a fuel cell to electrochemically convert chemical energy into an electric current.

In 2008, approximately 70% of the electricity used in the United States was generated by burning fossil fuels (coal, natural gas, petroleum liquids) (U.S. EIA 2010). The combustion of a fossil fuel to generate electricity can be either: 1) in a steam generating unit (also referred to simply as a “boiler”) to feed a steam turbine that, in turn, spins an electric generator: or 2) in a combustion turbine or a reciprocating internal combustion engine that directly drives the generator. Some modern power plants use a “combined cycle” electric power generation process, in which a gaseous or liquid fuel is burned in a combustion turbine that both drives electrical generators and provides heat to produce steam in a heat recovery steam generator (HRSG). The steam produced by the HRSG is then fed to a steam turbine that drives a second electric generator. The combination of using the energy released by burning a fuel to drive both a combustion turbine generator set and a stream turbine generator significantly increases the overall efficiency of the electric power generation process.

Coal is the most abundant fossil fuel in the United States and is predominately used for electric power generation. In 2008, approximately 49% of the net electricity generated in the U.S. was produced by coal (U.S. EIA 2010). Historically, electric utilities have burned solid coal in steam generating units. However, coal can also be first gasified and then burned as a gaseous fuel. The integration of coal gasification technologies with the combined cycle electric generation process is called an integrated gasification combined cycle (IGCC) system or a “coal gasification facility”. For the remainder of this document, the term “electric generating unit” or “EGU” is used to mean a solid fuel-fired steam generating unit that serves a generator that produces electricity for sale to the electric grid.

2. Coal-Fired Electric Generating Units

This section provides a summary overview of the types or ranks of coal that are typically burned in EGUs operating in the United States, the most commonly used combustion processes, and the resulting emissions of greenhouse gases.

2.1. Coals Burned in U.S. EGUs

In the United States, coals are ranked based on the degree of metamorphism (effectively, the geological age of the coal and the conditions under which the coal formed). These classification criteria have been standardized by the American Society for Testing and Materials (ASTM) method D-388. Under the ASTM method, coals are divided into four major categories called "ranks:" anthracite, bituminous coal, subbituminous coal, and lignite. Typical coal characteristics for the three most commonly used coal ranks are summarized in Exhibit 2-1.

Exhibit 2-1. Selected characteristics of major coal ranks used for electricity generation in the United States

Coal Rank[a]	Higher Heating Value (HHV) Range Defined by ASTM D-388	Typical Coal Moisture Content[b]	Coal Delivered for U.S. Electric Power Production in 2008[c,d]		
			Total Coal Quantity Delivered Nationwide	Average Ash Content	Average Sulfur Content
Bituminous	>10,500 Btu/lb	2 to 16%	463,943	10.6%	1.68%
Subbituminous	<10,500 Btu/lb and	15 to 30%	522,228	5.8%	0.34%
Lignite	< 8,300 Btu/lb	25 to 40%	68,945	13.8%	0.86%

[a] Anthracite coal use is limited to reclaiming coal from coal refuse piles for use in a few power plants located close to the anthracite mines in eastern Pennsylvania.

[b] Reference: U.S. EPA, 2001.

[c] Reference: U.S. EIA, 2010, Table 3.6.

[d] Includes data collected from electric utilities, independent power producers, and combined heat and power producers.

Most coal-fired EGUs in the United States burn either bituminous or subbituminous coals. Approximately one half of the tonnage of coals delivered to U.S. electric power generation facilities was subbituminous (49.5%), and another 44% was bituminous coal. Some coal-fired EGUs burn multiple coal ranks. At many of these facilities, the coals are blended together before firing. However, some facilities may switch between coal ranks because of site-specific considerations. The largest sources of bituminous coals burned in EGUs are mines in regions along the Appalachian Mountains, in southern

Illinois, and in Indiana. Additional bituminous coals are supplied from mines in Utah and Colorado. The vast majority of subbituminous coals are supplied from mines in Wyoming and Montana, and many EGUs burn subbituminous coals from the Powder River Basin (PRB) region in Wyoming. This material is often referred to simply as "PRB coal."

In general, the burning of lignite or anthracite by electric utilities is limited to those EGUs that are located near the mines supplying the coal. Lignite accounted for approximately 6.5% of the total tonnage of coal delivered to electric utility power plants in 2008. All of those facilities were located near the coal deposits from which the lignite was mined in Texas, Louisiana, Mississippi, Montana, or North Dakota. Similarly, anthracite use was limited to a few power plants located close to the anthracite mines in eastern Pennsylvania. The coal-fired EGUs at those facilities primarily burn anthracite that has been reclaimed from coal refuse piles of previous mining operations. In general, "coal refuse" means any by-product of coal mining or coal cleaning operations with an ash content greater than 50 % (by weight) and a heating value less than 13,900 kilojoules per kilogram (kJ/kg) (6,000 Btu per pound (Btu/lb) on a dry basis. Coal refuse piles from previous mining operations are primarily located in Pennsylvania and West Virginia. Current mining operations generate less coal refuse than older ones.

2.2. Coal Utilization in U.S. EGUs

Steam turbine power plants operate on the Rankine thermodynamic cycle. The steam is produced by the boiler, where water pumped into the boiler ("feedwater") passes through a series of tubes to capture heat released by coal combustion and then boils under high pressure to become superheated steam. The superheated steam leaving the boiler then enters the steam turbine throttle, where it powers the turbine and connected generator to make electricity.

After thc steam expands through the turbine, it exits the back end of the turbine into the surface condenser, where it is cooled and condensed back to water. This condensate is then returned to the boiler through high-pressure feed pumps for reuse. Heat from the condensing steam is normally rejected to cooling water circulated through the condenser which then goes to a surface water body, such as a river, or to an on-site cooling tower.

An EGU can be classified as either dry or wet bottom, depending on the ash removal technique used. Dry bottom boilers fire coals with high ash fusion temperatures, allowing for solid ash removal. In the less common wet bottom (slag tap) boilers, coal with a low ash fusion temperature is fired, and molten ash is drained from the bottom of the boiler.

Coal-fired EGUs use one of five basic coal utilization processes.

- Stoker-fired
- Pulverized coal (PC)
- Cyclone-fired
- Fluidized-bed combustion (FBC)
- Coal gasification (IGCC)

To improve the overall thermal conversion efficiency of the Rankine cycle, the majority of EGUs include a series of heat recovery sections. These sections are located downstream from the furnace chamber and are used to extract additional heat from the flue gas. The first section contains a "superheater," which is used to increase the steam temperature. The second heat recovery section contains a "reheater," which reheats the steam exhausted from the first stage of the steam turbine. This steam is then returned for another pass thorough a second stage of the turbine. The reheater is followed by an "economizer," which preheats the condensed feedwater recycled back to the boiler tubes in the furnace. The final heat recovery section is the "air heater," which preheats the ambient air used for coal combustion. The flue gas exhausted from the boiler passes through particulate matter (PM) and other air emissions control equipment before being vented to the atmosphere through a stack.

2.2.1. Stoker-Fired Coal Combustion

First introduced to the electric utility industry in the late 1800s, stoker-fired coal combustion is the oldest boiler coal-firing design. In a stoker-fired boiler, the coal is crushed and burned on a grate. Heated air passes upward through openings in the grate. Stokers are classified according to the way coal is fed to the grate – as underfeed stokers, overfeed stokers, and spreader stokers (see Exhibit 2-2). Stoker firing coal combustion is an obsolete technology for new coal-fired EGUs because the other newer coal combustion technologies provide superior coal combustion efficiency, applicability, and other advantages. There are still a few small stoker-fired EGUs in service in the U.S., but as these units are retired no new coal-fired stoker-fired EGUs are expected to be built. The majority of new stoker-fired boiler capacity is expected to occur at municipal solid waste combustor facilities and facilities burning solid biomass.

2.2.2. Pulverized-Coal Combustion

Pulverizing coal into a very fine powder allows the coal to be burned more easily and efficiently. For a PC-fired EGU, the coal must first be pulverized in a mill to the consistency of talcum powder (i.e., at least 70% of the particles will pass through a 200-mesh sieve). The pulverized coal is generally entrained in primary combustion air before being blown through the burners into the combustion chamber where it is fired in suspension. PC-fired boilers are classified by the firing position of the burners either as wall-fired or tangential-fired (see Exhibit 2-2).

A PC-fired boiler consists of multiple sections, and Exhibit 2-3 presents a simplified schematic of the major components of a PC-fired boiler using subcritical steam conditions. The pulverized coal is ignited and burned in the section of the boiler called the "furnace chamber" (or sometimes the "firebox"). Ambient air blown into the furnace chamber provides the oxygen required for combustion. The walls of the furnace chamber are lined with vertical tubes containing the feedwater. Heat transfer from the hot combustion gases in the furnace boils the water in the tubes to produce the high-temperature, high-pressure steam. The steam is separated from boiler water in a steam drum and sent to the steam turbine. The remaining water in the drum re-enters the boiler for further conversion to steam. The hot combustion products are vented from the furnace in a gas stream called collectively flue gas.

2.2.3. Cyclone Coal Combustion

Cyclone coal combustion technology was developed as an alternative to PC-firing because it requires less pre-processing of the coal and allows for the burning of lower rank coals with higher moisture and ash contents. Cyclone boilers use burner design and placement (i.e., several water-cooled horizontal burners) to produce high-temperature flames that circulate in a cyclonic pattern. The coal is crushed to a 4-mesh size, and then fed tangentially with primary air, to a horizontal cylindrical combustion chamber. In this chamber, small coal particles are burned in suspension, while the larger particles are forced against the outer wall.

Exhibit 2-2. Characteristics of coal-firing configurations used for U.S. EGUs

Coal-firing Configuration	Application to U.S. EGUs	Coal Combustion Process Description	Distinctive Design/Operating Characteristics	
Stoker-fired	● Oldest coal-firing design first introduced to the electric utility industry in the late 1800s. ● Not a significant contributor to overall U.S. nationwide MW generating capacity. ● New EGUs are not expected to use this coal-firing design because of the superior performance and advantages of newer coal combustion technologies.	Coal is crushed into large lumps and burned in a fuel bed on a moving, vibrating, or stationary grate. Coal is pushed, dropped, or thrown onto the grate by a mechanical device called a "stoker."	Spreader-stoker	A flipping mechanism throws the coal into the furnace above the grate. The fine coal particles burn in suspension while heavier coal lumps fall to the grate and burn in a fuel
			Underfeed	Coal fed by pushing the coal up underneath the burning
			Traveling grate	Coal is fed by gravity onto a moving grate and leveled by a stationary bar at the furnace entrance.
Pulverized-Coal Combustion	● Coal-firing design predominately used at existing U.S. EGUs ● In 2008, consumed ~ 92% of total coal consumed by U.S. EGUs.[a] ● Currently coal-firing design of choice for new large coal-fired EGUs (> 400 MWe) built in U.S.	Coal is ground to a fine powder that is pneumatically fed to a burner where it is mixed with combustion air and then blown into the furnace. The pulverized- coal particles burn in suspension in the furnace. Unburned and partially burned coal	Wall-fired	An array of burners fire into the furnace horizontally, and can be positioned on one wall or opposing walls depending on the furnace design.
			Tangential-fired (Corner-fired)	Multiple burners are positioned in opposite corners of the furnace producing a fireball that moves in a cyclonic motion and expands

Exhibit 2-2. (Continued)

Coal-firing Configuration	Application to U.S. EGUs	Coal Combustion Process Description	Distinctive Design/Operating Characteristics	
Cyclone	• Existing cyclone EGUs in U.S. constructed prior to 1981. • In 2008, consumed ~ 6% of total coal consumed by U.S. EGUs. • New EGUs are not expected to use this boiler type because of the commercial availability of FBC technology.	Coal is crushed into small pieces and fed through a burner into the cyclone furnace. A portion of the combustion air enters the burner tangentially creating a whirling motion to the incoming coal.	Designed to burn coals with low-ash fusion temperatures that are difficult to burn in PC boilers. The majority of the ash is retained in the form of a molten slag.	
Fluidized-bed Combustion	• FBC EGUs increasingly being built in U.S. to burn low rank coals, coal refuse, and blends of coal with other solid fuels such as petroleum coke or biomass. • In 2008, consumed approximately 2% of total coal consumed by U.S. EGUs.a	Coal is crushed into fine particles. The coal particles are suspended in a fluidized bed by upward-blowing jets of air. The result is a turbulent mixing of combustion air with the coal particles. Typically, the coal is mixed with a	Bubbling fluidized bed (BFB)	Operates at relatively low gas stream velocities and with coarse-bed size particles. Air in excess of that required to fluidize the bed passes through the bed in the form of bubbles.

Coal-firing Configuration	Application to U.S. EGUs	Coal Combustion Process Description	Distinctive Design/Operating Characteristics	
	● Atmospheric FBC EGUs are currently operating in the U.S. with generating capacities in the range of 250 to 300 MWe. ● No Pressurized FBC boilers currently used for U.S. EGUs	sorbent such as limestone (for SO_2 emission control). The unit can be designed for combustion within the bed to occur at atmospheric or elevated pressures. Operating temperatures for FBC are in the range of 1,500 to 1,650□F (800 to 900°C).	Circulating fluidized bed (CFB)	Operates at higher gas stream velocities and with finer-bed size particles. No defined bed surface. Must use high-volume, hot cyclone separators to recirculate entrained solid particles in flue gas to maintain the bed and achieve high combustion efficiency.
Coal Gasification (e.g., IGCC)	● Limited application to EGUs to date. ● Some new proposed EGU projects using coal gasification as part of IGCC plant.	Synthetic combustible gas ("syngas") derived from an on-site coal gasification process is burned in a combustion turbine. The hot exhaust gases from the combustion turbine pass through a heat recovery steam generator to produce steam for driving a steam turbine/generator unit.	Coal gasification units are unique from the other coal-firing configurations because a gaseous fuel (synfuel or syngas) is burned instead of solid coal and combines the Rankine and Brayton thermodynamic cycles as is the case for a combined cycle power plant.	

[a] Source: U.S. EIA, 2008.

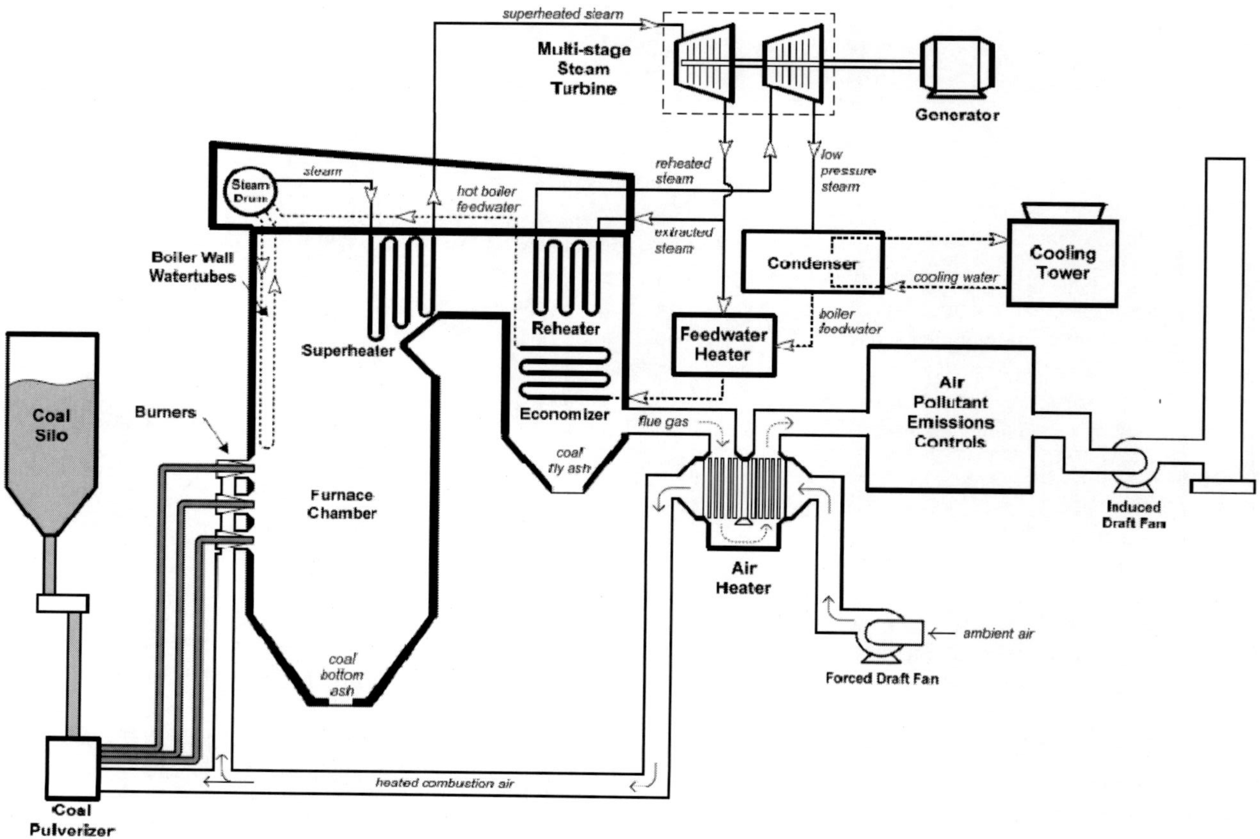

Exhibit 2-3. Simplified schematic of a PC-fired EGU using a subcritical boiler.

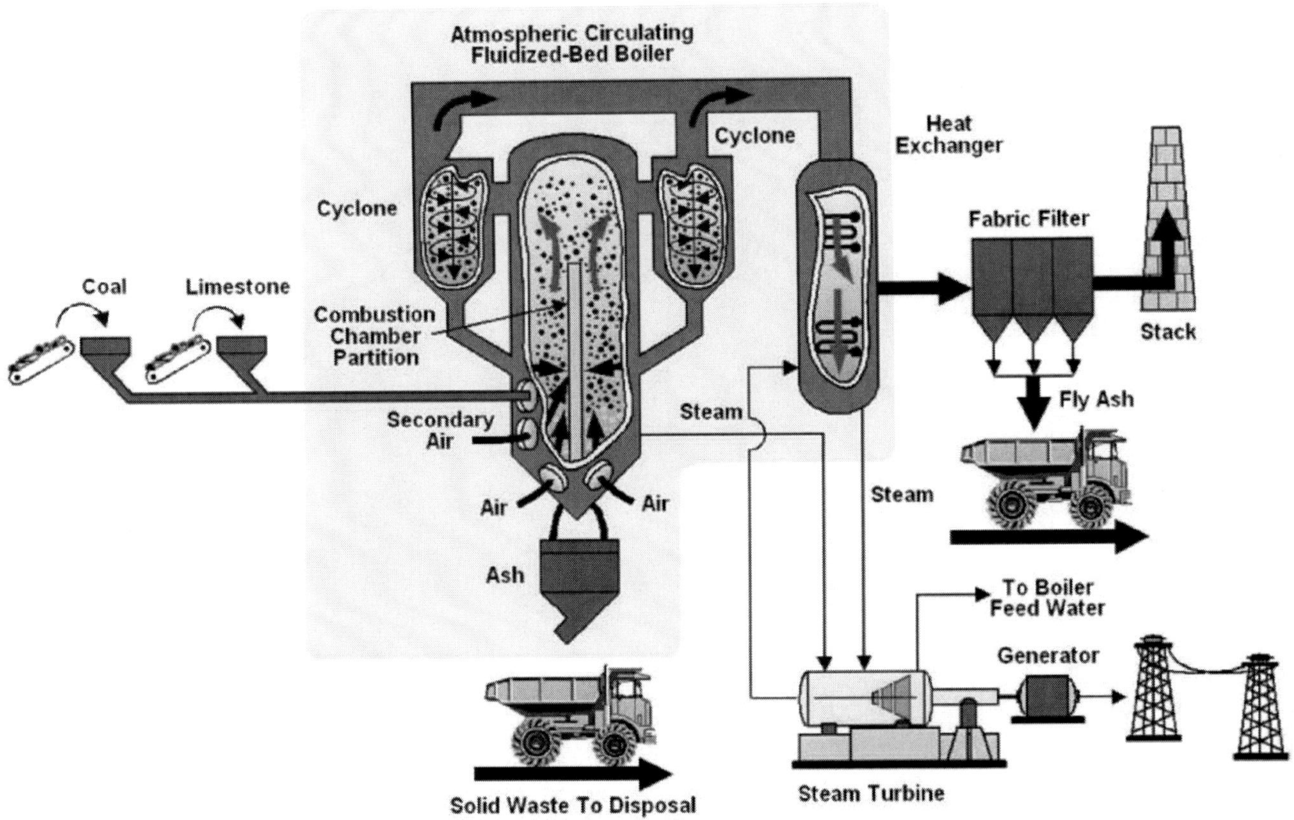

Source: NETL, 2010b, CCPI/Clean Coal Demonstrations Nucla CFB Demonstration Project, Project Fact Sheet.

Exhibit 2-4. Simplified schematic of an atmospheric circulating fluidized-bed (CFB) boiler power plant.

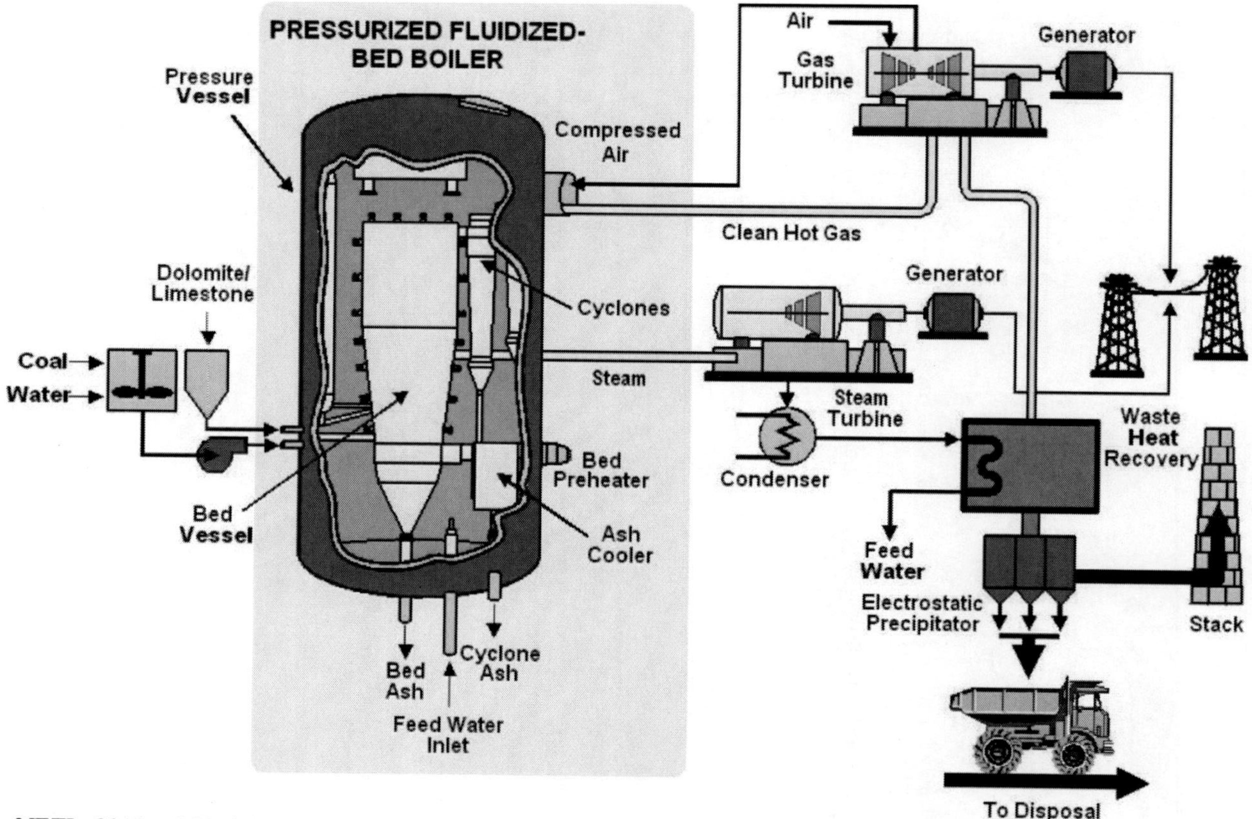

Source: NETL, 2010c, CCPI/Clean Coal Demonstrations Tidd PFBC Demonstration Project, Project Fact Sheet.

Exhibit 2-5. Simplified schematic of a pressurized fluidized-bed combustion (PFBC) power plant.

The high temperatures developed in the relatively small boiler volume, combined with the low fusion temperature of the coal ash, causes the ash to form a molten slag, which is drained from the bottom of the boiler through a slag tap opening. Existing cyclone EGUs in the U.S. were designed or installed before 1981. Cyclone EGUs have high nitrogen oxides (NOx) emission rates and no new cyclone boilers are expected to be built. Fluidized-bed combustion is an alternative technology that is able to burn lower rank coals without high NOx emissions.

2.2.4. Fluidized-Bed Combustion

The term "fluidized" refers to the state of the bed materials (fuel and inert material [or sorbent]) as gas passes through the bed. In a typical FBC EGU, combustion occurs when coal and a sorbent, such as limestone, are suspended through the action of primary combustion air distributed below the combustor floor. The gas cushion between the solids allows the particles to move freely, giving the bed a liquid-like characteristic (i.e., fluidized). FBC can occur in either atmospheric or pressurized boilers. Two fluidized bed designs can be used for atmospheric and pressurized FBC boilers: a bubbling fluidized bed or a circulating fluidized bed (CFB) (see Exhibit 2-2). An advantage of CFB boiler EGUs compared to PC-fired EGUs is fuel flexibility. A CFB boiler EGU can burn any rank of coal (including coal refuse), petroleum coke (a carbonaceous solid derived from oil refinery coker units or other cracking processes), and biomass without significant modifications.

The combustion temperature of a FBC boiler (1,500 to 1,650°F) is significantly lower than a PC-fired boiler (2,450 to 2,750°F), which results in lower NOX formation and the ability to capture sulfur dioxide (SO_2) with limestone injection in the furnace. Even though the combustion temperature of a FBC boiler is low, the circulation of hot particles provides efficient heat transfer to the furnace walls and allows longer residence time for carbon combustion and limestone reaction. This results in good combustion efficiencies, comparable to PC-fired EGUs.

Atmospheric CFB boilers have successfully been scaled-up and are operating at a number of facilities throughout the world. Exhibit 2-4 presents a simplified schematic of the major components of a CFB boiler EGU. Calcium in the sorbent combines with SO_2 gas to form calcium sulfite and sulfate solids, and solids exit the combustion chamber and flow into a hot cyclone. The cyclone separates the solids from the gases, and the solids are recycled for combustor temperature control. Heat in the flue gas exiting the hot cyclone is recovered in a series of heat recovery sections of the boiler to produce steam.

The superheated steam leaving the boiler then enters the steam turbine, which powers a generator to produce electricity. Like PC-fired EGUs, CFB boilers can be used with either subcritical or supercritical steam cycles.

Currently, the capacity of CFB subcritical boilers ranges from 25 to 350 MWe. Examples of these systems include (Foster Wheeler North America Corp., 2009):

- Two 300 MWe CFB subcritical boilers at the Jacksonville Energy Authority power plant in Jacksonville, Florida. These units are capable of burning either 100% coal or 100% petroleum coke or any combination of the two.
- Three 262 MWe CFB subcritical boilers at the Turow power plant in Poland. The fuel for these boilers is lignite with moisture content of 45% by weight.

The largest atmospheric CFB boiler in operation to date is a 460 MWe unit at a power plant owned by the Polish utility company Południowy Koncern Energetyczny SA (PKE) in Lagisza, Poland (Foster Wheeler North America Corp., 2009). This unit is also the world's first supercritical CFB boiler. The primary fuel burned in the unit is Polish bituminous coal. The commercial operation of this unit demonstrates the successful integration of CFB boiler technology with supercritical boiler technology. The unit features include a vertical evaporator with supercritical steam conditions (4,000 psia, 1,050/1,075°F) and a reported overall net plant efficiency of 41.6% (HHV basis). Based on the design and operating experience with the Lagisza Power Plant, both 600 and 800 MWe size supercritical CFB boiler designs with full commercial guarantees are being offered (Foster Wheeler North America Corp., 2009).

Pressurized fluidized-bed combustion (PFBC) systems are FBC systems that operate at elevated pressures (typically pressures of 1-1.5 MPa) and produce a high-pressure gas stream at temperatures that can drive a turbine. As with atmospheric FBC, two formats are possible, one with bubbling beds, the other with a circulating configuration. Currently, all operating units use bubbling beds. In a PFBC, the combustor and hot gas cyclones are all enclosed in a pressure vessel. Both coal and sorbent (for SO_2 emissions reductions) have to be fed across the pressure boundary, and similar provision for ash removal is necessary. For hard coal (i.e., bituminous coal) applications, the coal and limestone can be crushed together, and then fed as a paste, with 25% water. As with atmospheric FBC, a combustion temperature between 1,500 to

1,650°F (800 to 900°C) has the advantage of less NOX formation than in PC combustion. In addition, the effectiveness of a CCS system is increased due to the high pressure within the PFBC cycle and higher partial pressure of the CO_2 in the hot gas stream.

The initial or first-generation PFBC designs are based on directly burning crushed coal in the combustor. The high pressure gas is first expanded through a turbine and then heat is recovered from the turbine exhaust in a HRSG to produce steam, which is used to drive a conventional steam turbine. Exhibit 2-5 presents a simplified schematic of the major components of a PFBC EGU. A number of demonstration projects (ranging in size from 60 to 130 MWe) were conducted during the 1990s in Japan, Spain, Sweden, the United Kingdom, the U.S., and other countries. Japanese equipment manufacturers and electric power companies have led the commercial development of PFBC technology with the construction of several commercial-scale units.

- 360 MWe PFBC unit operated by Kyushu Electric Power Company at the Karita Power Station located near Kitakyushu, Japan. The unit began commercial operation in July 2001. The unit uses a supercritical boiler and has a reported net efficiency based on test results of 41.8% HHV (Asai, 2004).
- 250 MWe PFBC unit operated by Chugoku Electric Power Co., Inc. at Osaki Power Station located near Hiroshima, Japan. Unit 1 began commercial operation in 2000. The planned construction of a second PFBC unit at the facility was cancelled in 2008

More advanced second-generation PFBC system designs use a pressurized carbonizer to first process the feed coal into fuel gas and char (solid material that remains after light gases and tar have been driven-out during the initial stage of combustion). The PFBC burns the char to produce steam and to heat combustion air for the combustion turbine. The fuel gas from the carbonizer burns in a topping combustor linked to a combustion turbine, heating the gases to the rated firing temperature of the combustion turbine. Heat is recovered from the combustion turbine exhaust in a HRSG to produce steam, which is used to drive a conventional steam turbine. These systems are also called advanced circulating pressurized fluidized-bed combustion (APFBC) combined cycle systems.

2.2.5. Coal Gasification

An IGCC power plant uses a coal gasification system to convert coal into a synthetic gas, which is then used as fuel in a combined cycle electric generation process. Coal is gasified by a process in which coal or a coal/water slurry is reacted at high temperature and pressure with oxygen (or air) and steam in a vessel referred to as a "gasifier" to produce a combustible gas composed of a mixture of carbon monoxide (CO) and hydrogen. This gas is often referred to as synthetic gas or syngas. Gasification processes have been developed using a variety of designs including moving bed, fluidized bed, entrained flow, and transport gasifiers. Coal gasification processes are offered by a number of companies with varying degrees of existing commercial application (NETL, 2010a). Exhibit 2-6 presents a simplified schematic of the major components of an IGCC power plant. The hot syngas can then be processed to remove sulfur compounds, mercury, and PM before it is used to fuel a combustion turbine generator to produce electricity. The heat in the exhaust gases from the combustion turbine is recovered to generate additional steam. This steam, along with the steam produced by the gasification process, then drives a steam turbine generator to produce additional electricity.

The efficiency of an IGCC power plant is comparable to the latest advanced PC-fired and CFB EGU designs using supercritical boilers. The advantages of using IGCC technology can include greater fuel flexibility (e.g., capability to use a wider variety of coal ranks), potential improved control of PM, SO_2 emissions, and other air pollutants, with the need for fewer post-combustion control devices (e.g., almost all of the sulfur and ash in the coal can be removed once the fuel is gasified and prior to combustion), generation of less solid waste requiring disposal, and reduced water consumption when compared to an EGU using a supercritical boiler (U.S. EPA, 2006). Disadvantages of using IGCC include additional plant complexity, higher construction costs, and poorer performance at high altitude locations when compared to an EGU using a supercritical boiler. However, IGCC power plants offer the potential for lower control costs of CO_2 emissions because the CO_2 in the syngas can be removed prior to combustion. Interest by U.S. electric utilities in building new IGCC power plants is increasing because of site-specific considerations and potential cost benefits for the technology.

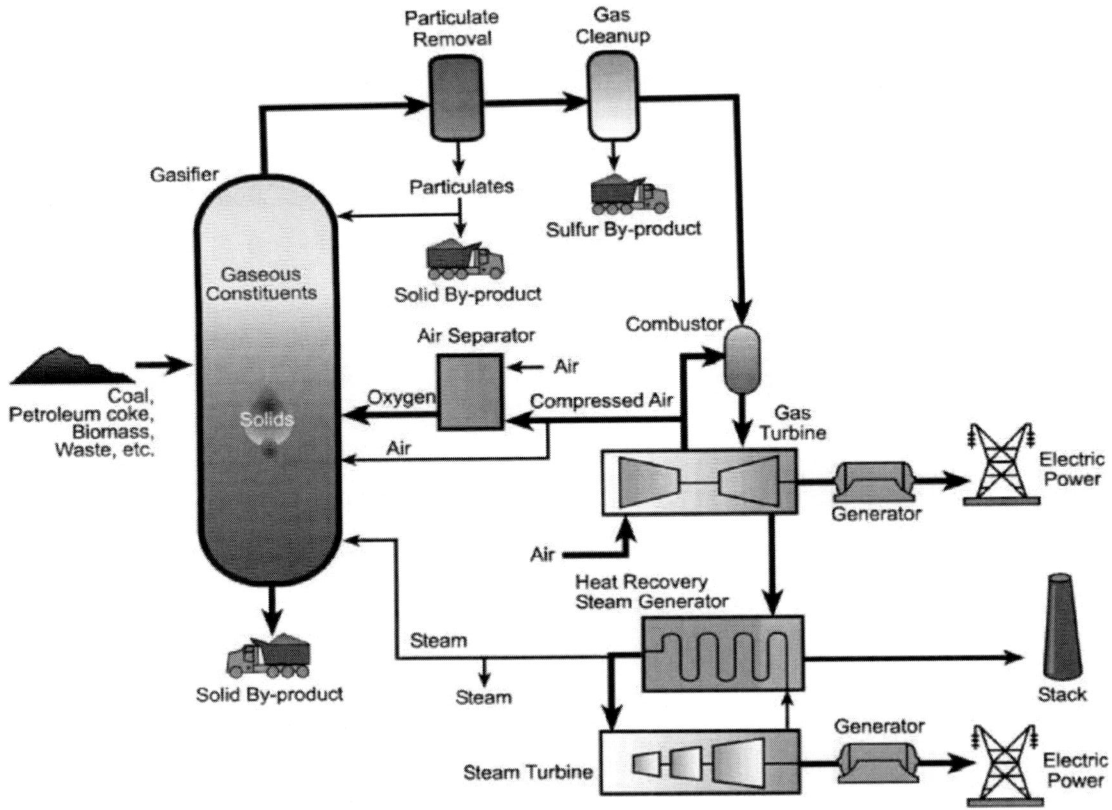

Source: NETL, 2010a, Overview of DOE's Gasification Program.

Exhibit 2-6. Simplified schematic of an integrated gasification combined cycle (IGCC) power plant.

Currently operating IGCC plants include the following U.S. and foreign plants (NETL, 2010a):

- 253 MWe IGCC plant at the NUON Willem-Alexander Power Plant in Buggenum, Netherlands. The unit began operation in 1994.
- 262 MWe IGCC plant at the Duke Energy Wabash River Power Station in Indiana. The unit began operation in 1995.
- 250 MWe IGCC plant at the Tampa Electric Company (TECO) Polk Power Station in Florida. The unit began operation in 1996.
- 400 MWe IGCC plant at the SUV power plant in Vresova, Czech Republic. The unit began operation in 1996.
- 283 MWe IGCC plant at the ELCOGAS power plant in Puertollano, Spain. The unit began operation in 1998.
- 250 MWe IGCC plant at the Joban Joint Electric Power Company Nakoso Power Station in Iwaki City, Japan. The unit began operation in 2007.

Over the past 5 years, a number of larger IGCC power plant projects have been proposed by U.S. electric utility companies. Some of these IGCC projects have been indefinitely delayed or canceled because of economic and regulatory factors, such as escalating project investment costs beyond initial estimates and unresolved cost recovery issues with State public utility commissions. One commercial IGCC project currently under construction is a 630 MWe IGCC facility at the Duke Energy Edwardsport Power Station in Knox County, Indiana.

Syngas produced by coal gasification can not only be used as a fuel to generate electricity or steam but also as a basic chemical building block for a large number of petrochemical and refining products. Because of these multiple uses, future IGCC projects may include facilities that integrate electricity generation with the production of other industrial outputs such as chemical feedstocks for manufacturing operations or hydrogen fuel for vehicles.

2.3. GHG Emissions from Coal-Fired EGUs

The principal chemical constituents of coal are carbon, hydrogen, oxygen, nitrogen, sulfur, moisture, and incombustible mineral matter (i.e., ash). When coal is burned, the carbon and hydrogen are oxidized to form the primary combustion products of CO_2 and water. Other combustion products such as NOX, SO_2, CO, and PM are formed in varying amounts.

The principal GHGs that enter the atmosphere because of human activities are CO_2, nitrous oxide (N_2O), methane (CH4), hydrofluorocarbons (HFC's), perfluorocarbons (PCF's), and sulfur hexafluoride (SF6). Of these, CO_2 is by far the most abundant GHG emitted from power production by coal utilization. To optimize overall efficiency for a given EGU, the unit is operated under conditions such that nearly all of the fuel carbon is converted to CO_2 during the combustion process. Methane is emitted during the mining and transport of coal but is not a significant by-product of EGU coal combustion. Fluorinated gases are not formed by coal combustion. Sulfur hexafluoride might be used at the power plant switchyard, but the switchyard is not typically considered part of the EGU.

Formation of N_2O during the combustion process results from a complex series of reactions and its formation is dependent upon many factors. However, the formation of N_2O is minimized when combustion temperatures are kept high and excess air is kept to a minimum. PC-fired EGUs are typically operated at these conditions and are not significant sources of N2O emissions. However, FBC EGUs can have measurable N2O emissions, resulting from the lower combustion temperatures and the use of selective noncatalytic reduction (SNCR) to reduce NOX emissions. Operating factors impacting N2O formation include combustion temperature, excess air, and sorbent feed rate (Korhonen, 2001). The N2O formation resulting from SNCR depends upon the reagent used, the amount of reagent injected, and the injection temperature (Weijuan, 2007).

2.4. Factors Impacting Coal-Fired EGU CO_2 Emissions

The level of CO_2 emissions that can potentially be released from a given coal-fired EGU depends on the type of coal burned, the overall efficiency of the power generation process, and use of air pollution control devices.

2.4.1. Impact of Coal Rank on CO_2 Emissions from EGUs

The amount of CO_2 that potentially can be emitted from a coal-fired EGU varies depending on the coal rank burned. The amount of heat released by coal combustion depends on the amounts of carbon, hydrogen, and oxygen present in the coal and, to a lesser extent, on the sulfur content. Hence, the ratio of carbon to heat content depends on these heat-producing components of coal, and these components vary by coal rank. **Exhibit 2-7** presents a comparison of the CO_2 emissions for the average heating values of U.S. coals. The values presented in the table are arithmetic averages and assume complete combustion. Based on these averages, in general anthracite emits the largest amount of CO_2 per million Btu (MMBtu), followed by lignite, subbituminous coal, and bituminous coal. However, for a given coal rank there is variation in the CO_2 emission factor depending on the coal bed from which the coal is mined.

Exhibit 2-7. CO_2 emission factors for coal by coal rank

Coal Rank	CO_2 Emissions per Unit of Heat Input (lbs CO_2/MMBtu)	
	U.S. Average	Range Across States with Coal Rank Deposits
Anthracite	227.4	227.4
Bituminous	205.3	201.3 to 211.6
Subbituminous	211.9	207.1 to 214.0
Lignite	216.3	211.7 to 220.6

Source: U.S. EIA (Hong, R. and E. Slatick, 1994).

In addition to the lower CO_2 emissions rate per unit of heat input (lbs CO_2/MMBtu), due to the inherent moisture in subbituminous and lignite coals, all else being equal a bituminous coal-fired boiler is more efficient than a corresponding boiler burning subbituminous or lignite coal. Therefore, switching from a low to a high-rank coal will tend to lower GHG emissions from the utility stack. However, overall GHG emissions might not be lowered by switching to bituminous coal. All coal mining operations release coal bed methane to the atmosphere during the mining process. Some bituminous coal reserves release significant amounts of methane, which could, in theory, offset GHG savings. Additional factors when considering overall GHG emissions include the fuel needs to mine, process, and transport the coal.

Additional solid fuels burned in EGUs include petroleum coke, biomass, and municipal solid waste (MSW). Petroleum coke has one of the highest CO_2 emissions rate (225 lb CO_2/MMBtu) of commonly used solid fuels. MSW combustors provide significant GHG reductions as an alternate to landfills (Kaplan, 2008). However, due to the difficulties associated with transporting large amounts of solid waste, MSW combustor facilities used for electrical power generation are typically limited to less than 100 MW of electrical output.

Of the gaseous and liquid fossil fuels used in steam generating units, natural gas combustion releases approximately 117 pounds of CO_2 per MMBtu, distillate oil releases 161 lb CO_2/MMBtu, and residual oil releases 174 lb CO_2/MMBtu. However, none of these fuels are typically used in new baseload steam generating units (e.g., boilers). Natural gas and distillate oil are significantly more expensive per unit heat input than coal. In addition, combustion turbines burning natural gas and distillate oil generate power more efficiently than a boiler burning natural gas and distillate oil. New baseload electric generation based on the use of either natural gas or distillate would likely use combined cycle combustion turbines. Therefore, aside from small amounts of natural gas for startup, shutdown, and potentially for combustion control, few new steam generating units are expected to burn significant quantities of either of these fuels directly in the boiler. Existing EGUs that burn natural gas and distillate oil tend to be older units that operate in a peaking or cycling mode. However, several base load coal-fired EGUs have been converted to natural gas. Natural gas-fired boilers tend to be less efficient than coal-fired boilers; however, they can startup and change loads more quickly than similar coal-fired boilers, do not typically require post combustion controls, and the fuel handling is simpler. Residual oil also tends to be more expensive than coal per unit of heat input, and because post-combustion environmental controls would still often be required, it is also not a common fuel choice for EGUs in the Lower 48 States. There has not been a new residual oil-fired EGU built in the Lower 48 States since 1981.

2.4.2 Impact of Coal-Fired EGU Efficiency on CO_2 Emissions

As the thermal efficiency of a coal-fired EGU is increased, less coal is burned per kilowatt-hour (kWh) generated, and there is a corresponding decrease in CO_2 and other air emissions. There is no standardized procedure for continuous on-line measurement of coal-fired EGU thermal efficiency (Peltier, 2010). However, a near approximation performed under EPA's Acid Rain Program collects heat input and gross megawatt output on an hourly basis to calculate gross heat rate. The heat input is derived from standardized continuous emission monitors, while the utility supplies gross megawatt output. The electric energy output as a fraction of the fuel energy input expressed in percentage is a commonly-used practice for reporting the efficiency of a coal-fired EGU. The greater the output of electric energy for a given amount of fuel energy input, the higher the efficiency for the electric generation process. Heat rate is another common way to express efficiency. Heat rate is expressed as the number of Btu or kJ required to generate a kWh of electricity. Lower heat rates are associated with more efficient power generating plants. Although the same basic formula is used to calculate efficiency for coal-fired EGUs, there are different methodologies for measuring the appropriate parameters. For example, the varying accuracy of the different methodologies can cause discrepancies in the measurement the heating value of the coal burned.

Efficiency can be calculated using the higher heating value (HHV) or the lower heating value (LHV) determined for the fuel. The HHV is the heating value directly determined by calorimetric measurement of the fuel in the laboratory. The LHV is calculated using a formula to account for the moisture in the fuel (i.e., subtract the energy required to vaporize the water in the coal and is thus not available to produce steam) and is a smaller value than the HHV. Consequently, the HHV efficiency for a given EGU is always lower than the corresponding LHV efficiency, because the reported heat input is larger. For bituminous coals the HHV efficiency value is typically about 2 percentage points lower than the corresponding LHV efficiency. For higher moisture subbituminous coals and lignites, the HHV efficiency is approximately 3 to 5 percentage points lower than the corresponding LHV efficiency (depending on moisture content). In engineering practice, HHV is typically used in the U.S. to express the efficiency of steam electric power plants while in Europe the practice is to use LHV.

Similarly, the electric energy output for an EGU can be expressed as either of two measured values. One value relates to the amount of total electric power generated by the EGU, or "gross output." However, a portion of this electricity must be used by the EGU facility to operate the unit, including pumps, fans, electric motors, and pollution control equipment. This in-facility electrical load, often referred to as the "parasitic load," reduces the amount of power that can be delivered to the transmission grid for distribution and sale to customers. Consequently, electric energy output is also expressed in terms of "net output," which reflects the EGU gross output minus its parasitic load.

When using efficiency to compare the effectiveness of different coal-fired EGU configurations and the applicable GHG emissions control technologies, it is important to ensure that all efficiencies are calculated using the same type of heating value (i.e., HHV or LHV) and the same type of electric energy output (i.e., gross MWh or net MWh).

Although there is a direct inverse correlation between coal-fired EGU efficiency and CO_2 emissions, other factors must be considered when comparing the effectiveness of GHG control technologies to improve the efficiency of a given coal-fired EGU. The actual overall efficiency that a given coal-fired EGU achieves is determined by the interaction of a combination of site-specific factors that impact efficiency to varying degrees. These factors include:

- *EGU thermodynamic cycle* – EGU efficiency can be significant improved by using a supercritical or ultra-supercritical steam cycle.
- *EGU coal rank and quality* – EGUs burning higher quality coals (e.g., bituminous) tend to be more efficient than EGUs burning lower quality coals (e.g., lignite).
- *EGU plant size* – The electric-generating capacity of EGUs ranges from approximately 25 to 1,300 MWe. Assuming an EGU efficiency of 33% (a typical efficiency for existing coal-fired EGUs), this corresponds to a heat input range of 250 to 13,400 MMBtu/hr. EGU efficiency generally increases with size because the boiler and steam turbine losses are lower for larger equipment. However, as equipment size increases the differences in these losses start to taper off.
- *EGU pollution control systems* – The electric power consumed by air pollution control equipment reduces the overall efficiency of the EGU.
- *EGU operating and maintenance practices* – The specific practices used by an individual electric utility company for combustion optimization, equipment maintenance, etc. can affect EGU efficiency.
- *EGU cooling system* – The temperature of the cooling water entering the condenser can have impacts on steam turbine performance. Once-through cooling systems can have an efficiency advantage over recirculating cooling systems (e.g., cooling towers). However, once-though cooling systems typically have larger water related ecological concerns than recirculating cooling systems.
- *EGU geographic location* – The elevation and seasonal ambient temperatures at the facility site potentially may have a measureable impact on EGU efficiency. At higher elevations, air pressure is lower and less oxygen is available for combustion per unit volume of ambient air than at lower elevations. Cooler ambient temperatures theoretically could increase the overall EGU efficiency by increasing the draft pressure of the boiler flue gases and the condenser vacuum, and by increasing the efficiency of a condenser recirculating cooling system.

- *EGU load generation flexibility requirements* – Operating an EGU as a baseload unit is more efficient than operating an EGU as a load cycling unit to respond to fluctuations in customer electricity demand.
- *EGU equipment manufacturers* – The efficiency specifications of major EGU components such as boilers, turbines, and electrical generators provided by equipment manufacturers can affect EGU efficiency.
- *EGU plant components* – EGUs using the optimum number of feedwater heaters, high-efficiency electric motors, variable speed drives, better materials for heat exchangers, etc. tend to be more efficient.

Because of these factors, coal-fired EGUs that are identical in design but operated by different utility companies in different locations may have different efficiencies. Thus, the level of effectiveness of a given GHG control technology used to improve the efficiency at one coal-fired EGU facility may not necessarily directly transfer to a coal-fired EGU facility at a different location.

2.4.3. Impact of SO_2 Controls on Coal-Fired EGU CO_2 Emissions

The SO_2 emissions from new coal-fired EGUs, or retrofitting of an existing facility without specific SO_2 controls, are controlled using flue gas desulfurization (FGD) technology to remove the SO_2 before it is vented to the atmosphere. The selection of the type of FGD technology will impact overall GHG emissions. All FGD processes require varying amounts of electric power to operate, which contributes to the overall parasitic load of the unit. The FGD parasitic load requirements are typically between 1-2% of the gross output of the facility. In addition, some FGD processes use carbon-containing reagents (e.g., carbonates) that form CO_2 as a byproduct of the chemical reactions of the reagent with SO_2.

For a typical unit, the CO_2 that is chemically created in a scrubber adds an additional 1% to the overall GHG emissions, but it can be as high as 3% for facilities burning high sulfur coals. However, from an overall GHG emissions standpoint the use of FGD technologies that do not form byproduct CO_2, such as lime-based scrubbers, do not necessarily reduce emissions. Lime is manufactured by heating limestone in the absence of oxygen to remove a molecule of CO_2 ($CaCO_3$ + heat $\rightarrow$ $CaO + CO_2$).

Exhibit 2-8. CO_2 formation from coal-fired EGU flue gas desulfurization (FGD) processes

FGD Type	Reagent	Forms CO_2	Overall Reaction(s)	Reference
Wet Scrubbing	Limestone ($CaCO_3$)	yes	$CaCO_3 + SO_2 + \frac{1}{2}H_2O \rightarrow CaSO_3 \bullet \frac{1}{2}H_2O(s) + CO_2$; $CaCO_3 + SO_2 + 2H_2O + \frac{1}{2}O_2 \rightarrow CaSO_4 \bullet 2H_2O(s) + CO_2$	Ref 1
	Magnesium-enhanced lime; Dolomitic lime [$Ca(OH)_2 \bullet Mg(OH)_2$]	no	$10Ca(OH)_2 + 11SO_2 + Mg(OH)_2 \rightarrow 10CaSO_3 \bullet \frac{1}{2}H_2O(s) + MgSO_3 + 6H_2O$	Ref 1,4
	Dual Alkali; Sodium solution and lime	no	$2NaOH + SO_2 \rightarrow Na_2SO_3 + H_20$; $H_2O + Na_2SO_3 + SO_2 \rightarrow 2NaHSO_3$; $2NaHSO_3 + Ca(OH)_2 \rightarrow Na_2SO_3 + CaSO_3 \bullet \frac{1}{2}H_2O + 3/2H_2O$; $Na_2SO_3 + Ca(OH)_2 \rightarrow 2NaOH + CaSO_3$	Ref 2
	Dual Alkali (Dowa)	yes	$Al_2O_3 \bullet Al_2(SO4)_3 + 3SO_2 + 3/2O_2 \rightarrow 2Al_2(SO4)_3$; $2Al_2(SO4)_3 + 3CaCO_3 \rightarrow Al_2O_3 \bullet Al_2(SO4)_3 + 2CaSO_4(s) + 3CO_2$	Ref 3
	Seawater	yes	$2NaHCO_3 + SO_2 \rightarrow Na_2SO_3 + 2CO_2 + H_20$; $Na_2SO_3 + \frac{1}{2}O_2 \rightarrow Na_2SO_4$	Ref 3
	Magnesium oxide (MgO)	no	$MgO + SO_2 \rightarrow MgSO_3$	Ref 7
	Hydrogen Peroxide (H_2O_2)	no	$H_2O_2 + SO_2 \rightarrow H_2SO_4$	Ref 3
	Sodium hydroxide (NaOH)	no	$2NaOH + SO_2 \rightarrow Na_2SO_3 + H_20$; $Na_2SO_3 + \frac{1}{2}O_2 \rightarrow Na_2SO_4$	Ref 3

Exhibit 2-8. (Continued)

FGD Type	Reagent	Forms CO_2	Overall Reaction(s)	Reference
Dry/Semi-dry Scrubbing	Hydrated calcitic lime ($Ca(OH)_2$)	no	$Ca(OH)_2 + SO_2 \rightarrow CaSO_3 \bullet ½H_2O(s) + ½H_2O$; $Ca(OH)_2 + SO_2 + H_2O + ½O_2 \rightarrow CaSO_4 \bullet 2H_2O(s)$	Ref 1
	Sodium bicarbonate ($NaHCO_3$)	yes	$2NaHCO_3 + SO_2 \rightarrow Na_2SO_3 + 2CO_2 + H_20$; $2NaHCO_3 + SO_2 + ½O_2 \rightarrow Na_2SO_4 + 2CO_2 + H_20$	Ref 8
	Sodium sesquicarbonate (trona)	yes	$2(Na_2CO_3 \bullet NaHCO_3 \bullet 2H_2O) + 3SO_2 \rightarrow 3Na_2SO_3 + 5H_2O + 4CO_2$; $2(Na_2CO_3 \bullet NaHCO_3 \bullet 2H_2O) + 3SO_2 + 3/2O_2 \rightarrow 3Na_2SO_4 + 5H_2O + 4CO_2$	Ref 8
	Sodium carbonate (soda ash, Na_2CO_3)	yes	$Na_2CO_3 + SO_2 + ½O_2 \rightarrow Na_2SO_4 + CO_2$	Ref 8
	Pulverized limestone	yes	$CaCO_3 + SO_2 + 2H_2O + ½O_2 \rightarrow CaSO_4 \bullet 2H_2O(s) + CO_2$	Ref 1
Other Processes	Ammonia (NH_3)	no	$SO_2 + 2NH_3 + H_2O + ½O_2 \rightarrow (NH_4)2SO_4$	Ref 1,5
	Activated carbon	no	$SO_2 + H_2O + ½O_2 \rightarrow H_2SO_4$	Ref 6

Ref 1. Srivastava, R., W. Jozewicz, and C. Singer, 2001.

Ref 2. Srivastava, R. and W. Jozewicz , 2001.

Ref 3. Davenport, 2006. Ref 4. Benson, 2003. Ref 5. He, 2002.

Ref 6. EPA, 2005. Ref 7. Shand, 2009. Ref 8. Maziuk, 2002.

Unless the CO_2 is sequestered at the lime production facility, overall GHG emissions will be similar. A list of FGD processes used for controlling SO_2 emissions from coal-fired EGUs is presented in Exhibit 2-8, identifying those processes that chemically form additional CO_2.

3. Coal-Fired EGU CO_2 Control Technologies

The development of effective and commercially viable CO_2 control technologies for coal-fired EGUs is receiving widespread attention from the utilities, technology providers, and government agencies. Some CO_2 control technologies are still in the research and development phase and are not yet ready for commercial application. Other CO_2 control technologies are being demonstrated at larger scales and are progressing towards commercial viability. This remains an active area of research and new projects and technology advances are reported routinely. The discussions of CO_2 mitigation technologies and options presented in this section are based on the development status of a given technology as described in publicly available information as of May 2010.

3.1. Coal-Fired EGU CO_2 Emissions Control Approaches

A number of technologies lowering CO_2 emissions from coal-fired EGUs are currently commercially available or under development. These control measures use one of two basic approaches to reduce the amount of CO_2 released to the atmosphere: 1) by reducing the amount of fuel used (and the amount of CO_2 formed) by improving the energy efficiency of the electrical generation process, or 2) by separating the CO_2 for long-term storage using carbon capture technology.

3.1.1. Efficiency Improvements

When the efficiency of the power generation process is increased, less coal is burned to produce the same amount of electricity. This provides the benefits of lower fuel costs and reduced air pollutant emissions (including CO_2). A number of energy efficiency technologies are available for application to both existing and new coal-fired EGU projects that can provide incremental step improvements to the overall thermal efficiency. The energy efficiency technologies with the potential to achieve the greatest improvements in electric power generation efficiency involve EGU design, equipment selection, and cost decisions that are typically incorporated during the planning and engineering design phases for a new EGU project.

3.1.2. Carbon Capture and Storage

Carbon capture and storage (CCS) involves the separation and capture of CO_2 from flue gas, or syngas in the case of IGCC. It also requires pressurization of the captured CO_2, transportation via pipeline if necessary, and injection and long-term geologic storage. Several different technologies, at varying stages of development, may be considered for the CO_2 separation and capture. Some have been demonstrated at the slip-stream or pilot-scale, while many others are still at the bench-top or laboratory stage of development.

Development of commercially viable processes for capturing CO_2 from EGUs is being funded by U.S. DOE, electric utility companies, and other organizations. These processes typically use solvents, solid sorbents, and membrane-based technologies for separating and capturing CO_2. Amine-based solvent systems are in commercial use for scrubbing CO_2 from industrial flue gases and process gases. However, solvents have yet to be applied to removing the large volumes of CO_2 that would be required for a coal-fired EGU. Solid sorbents can be used to capture CO_2 through chemical adsorption, physical adsorption, or a combination of the two effects. Membrane-based capture uses permeable or semi-permeable materials that allow for the selective transport/separation of CO_2. Oxy-combustion uses high-purity oxygen (O_2) instead of air to combust coal, producing a highly concentrated CO_2 stream that does not require a separation/capture step.

Once the CO_2 is captured, it is transported, if necessary, and stored. Geologic formations such as oil and gas reservoirs, unmineable coal seams, and underground saline formations are potential options for long-term storage. Basalt formations and organic rich shales are also being investigated for potential use as storage. Beneficial reuse (e.g., enhanced oil recovery or carbonation) is a potential alternative to strict storage that provides potential revenue to offset a portion of the CCS costs.

One recent study prepared for the U.S. DOE by the Pacific Northwest National Laboratory (PNNL, 2009) evaluated the development status of various CCS technologies. The study addressed the availability of capture processes; transportation options (CO_2 pipelines); injection technologies; and measurement, verification, and monitoring technologies. The study concluded that, in general, CCS is technically viable today. However, full-scale carbon separation and capture systems have not yet been installed and fully integrated at an EGU. The study also did not address the cost or energy requirements of implementing CCS technology. For up-to-date information on Department of Energy's National Energy Technology Laboratory's (NETL) Carbon Sequestration Program go to the NETL web site at: http://www.netl.doe.gov/technologies/carbon_seq/.

In 2010, an Interagency Task Force on Carbon Capture and Storage was established to develop a comprehensive and coordinated federal strategy to speed the commercial development and deployment of CCS technologies. The Task Force is specifically charged with proposing a plan to overcome the barriers to the widespread, cost-effective deployment of CCS within 10 years, with a goal of bringing 5 to 10 commercial demonstration projects online by 2016. As part of its work, the Task Force prepared a report that summarizes the state of CCS and identified technical and non-technical barriers to implementation. For additional information on the Task Force and its findings on CCS, go to: http://www.epa.gov/climatechange/policy task force.html. Because the development status of CCS technologies and their applicability to coal-fired EGUs are thoroughly discussed in the Task Force report, there will be no further discussion in this document.

3.2. Efficiency Improvements for Existing Coal-fired EGU Projects

Numerous efficiency improvements can be applied to coal-fired EGUs to increase thermal efficiency of power production (NETL, 2008, Sargent & Lundy, 2009; U.S. DOE, 2009; U.S. DOE, 2010). One specific example is the NETL study, which conducted a literature review of published articles and technical papers identifying potential efficiency improvement techniques applicable to existing coal-fired EGUs. Efficiency improvements can be expressed in different formats; they may be reported as an absolute change in overall efficiency (e.g., a change from 40% to 42% represents a 2% absolute increase). They may also be presented as the relative change in efficiency (e.g., a change from 40% to 42% is a relative change in efficiency and fuel use of 5%). The relative change in efficiency is the most consistent approach, since it corresponds to the same change in heat rate.

A summary of the findings from the NETL study is presented in **Exhibit 3-1**. The efficiency percentages were converted to a common basis so that all of the data could be compared. All of these improvements could not necessarily be implemented at each coal-fired EGU because of site-specific factors.

3.3. Efficiency Improvements for New Coal-Fired EGU Projects

3.3.1. Steam Cycle

The theoretical maximum achievable thermal efficiency achievable by an EGU using the Rankine cycle regardless of the technologies used is approximately 63% because of thermodynamic limitations and energy losses that cannot be recovered. Existing coal-fired EGUs using the Rankine cycle operate well below this limit. If the energy input to the cycle is kept constant, increasing the pressures and temperatures for the water-steam cycle will increase the output and the overall efficiency. However, a practical limitation to the higher pressure and temperatures that can be achieved in a boiler is the availability of boiler materials that can withstand these elevated conditions over an acceptable service life. The majority of existing PC-fired EGUs have subcritical boilers. Subcritical boilers typically operate at pressures of 2,400 psi (17 MPa) and at temperatures between 1,000 to 1,050°F (540 to 570°C). However, subcritical boilers can be designed to operate at steam pressures as high as 3,200 psi (22 MPa) and steam temperatures as high as 1,050°F (570°C).

The use of materials that can withstand the high-temperature and pressure of supercritical steam conditions allows for substantial improvements in efficiency for EGUs. "Supercritical" is a thermodynamic term describing the state of a substance where there is no clear distinction between the liquid and the gaseous phase (i.e., they are a homogenous fluid). Technically, the term "boiler" should not be used for a supercritical pressure steam generator, as no "boiling" actually occurs in this device, but it is common practice to use the term "Benson boiler." Supercritical EGUs typically use steam pressures of 3,500 psi (24 MPa) and steam temperatures of 1,075°F (580°C). However, supercritical boilers can be designed to operate at steam pressures as high as 3,600 psi (25 MPa) and steam temperatures as high as 1,100°F (590°C). Above this temperature and pressure the steam is sometimes called "ultra-supercritcal".

For a supercritical boiler, the feed water enters the boiler, is converted to steam, and is passed directly to the steam turbine (a supercritical boiler does not have a steam drum as shown in Exhibit 2-3 for a subcritical boiler PC-fired EGU configuration). Because the water-steam cycle medium is a single phase fluid with homogeneous properties, there is no need to separate steam from water in a drum. Supercritical boilers operate as once-through boilers in which the water and steam generated in the furnace waterwalls passes through only once. This eliminates the need for water/steam separation in drums during operation, and allows a simpler separator to be employed during start-up conditions. Because these units do not have thick-walled steam drums, their start-up times are quicker, further enhancing efficiency and plant economics. Due to the availability of steam turbines that are designed for supercritical steam conditions, supercritical applications are presently limited to facilities of approximately 200 MWe gross output or more. Supercritical boilers are a well-established technology, and over 500 supercritical plants are currently operating worldwide (VGB, 2008).

The majority of EGUs have a single reheat cycle where the steam is first passed through the high pressure portion of the steam turbine and is then reheated in the boiler prior to passing through the remainder of the turbine. This process increases the efficiency of the EGU without increasing the maximum steam temperature. An additional steam cycle improvement that further increases efficiency is the use of a double reheat cycle, which reduces fuel use by approximately 1.5% compared to a similar EGU using a single reheat cycle (Retzlaff, 1996). The efficiency benefits of using a double reheat cycle have been recognized since the 1960s. However, the additional cost of a double reheat cycle has made a single reheat cycle typical for the majority of EGUs in the U.S.

To establish performance and cost baselines for analyzing EGU technology, NETL funded an independent assessment of the cost and performance of fossil energy power systems. The assessment specifically includes PC-fired boilers, IGCC, and natural gas-fired combined cycle systems in a consistent technical and economic manner reflecting market conditions for plants starting operation in 2010 (NETL, 2007). Performance and cost estimates were prepared for each configuration with and without CO_2 CCS. The Total Plant Cost (TPC) and Operation and Maintenance (O&M) costs for each of the cases in the study were estimated in January 2007 dollars and assumed plant construction on a generic site. The costs do not include owner cost and additional costs for special site-specific considerations at a given site.

For the PC-fired systems, the NETL study included a comparison of subcritical (2,400 psig/1,050°F/1,050°F) and supercritical (3,500 psig/1,100°F /1,100°F) PC-fired EGUs, each rated at nominal 550 MWe net capacity and firing Illinois No. 6 bituminous coal. A summary comparing the results for the subcritical unit versus the supercritical unit (without CO_2 CCS) is presented in Exhibit 3-2. The analysis shows an efficiency (HHV) of 36.8% for the subcritical boiler compared to 39.1% for the supercritical boiler.

Exhibit 3-1. Existing coal-fired EGU efficiency improvements reported for actual efficiency improvement projects

Efficiency Improvement Technology	Description	Reported Efficiency Increase[a]
Combustion Control Optimization	Combustion controls adjust coal and air flow to optimize steam production for the steam turbine/generator set. However, combustion control for a coal-fired EGU is complex and impacts a number of important operating parameters including combustion efficiency, steam temperature, furnace slagging and fouling, and NOX formation. The technologies include instruments that measure carbon levels in ash, coal flow rates, air flow rates, CO levels, oxygen levels, slag deposits, and burner metrics as well as advanced coal nozzles and plasma assisted coal combustion.	0.15 to 0.84%
Cooling System Heat Loss Recovery	Recover a portion of the heat loss from the warm cooling water exiting the steam condenser prior to its circulation thorough a cooling tower or discharge to a water body. The identified technologies include replacing the cooling tower fill (heat transfer surface) and tuning the cooling tower and condenser.	0.2 to 1%
Flue Gas Heat Recovery	Flue gas exit temperature from the air preheater can range from 250 to 350°F depending on the acid dew point temperature of the flue gas, which is dependent on the concentration of vapor phase sulfuric acid and moisture. For power plants equipped with wet FGD systems, the flue gas is further cooled to approximately 125°F as it is sprayed with the FGD reagent slurry. However, it may be possible to recover some of this lost energy in the flue gas to preheat boiler feedwater via use of a condensing heat exchanger.	0.3 to 1.5%
Low-rank Coal Drying	Subbituminous and lignite coals contain relatively large amounts of moisture (15 to 40%) compared to bituminous coal (less than 10%). A significant amount of the heat released during combustion of low-rank coals is used to evaporate this moisture, rather than generate steam for the turbine. As a result, boiler efficiency is typically lower for plants burning low-rank coal. The technologies include using waste heat from the flue gas and/or cooling water systems to dry low-rank coal prior to combustion.	0.1 to 1.7%

Efficiency Improvement Technology	Description	Reported Efficiency Increase[a]
Sootblower Optimization	Sootblowers intermittently inject high velocity jets of steam or air to clean coal ash deposits from boiler tube surfaces in order to maintain adequate heat transfer. Proper control of the timing and intensity of individual sootblowers is important to maintain steam temperature and boiler efficiency. The identified technologies include intelligent or neural-network sootblowing (i.e., sootblowing in response to real-time conditions in the boiler) and detonation sootblowing.	0.1 to 0.65%
Steam Turbine Design	There are recoverable energy losses that result from the mechanical design or physical condition of the steam turbine. For example, steam turbine manufacturers have improved the design of turbine blades and steam seals which can increase both efficiency and output (i.e., steam turbine dense pack technology).	0.84 to 2.6

Source: NETL, 2008

[a] Reported efficiency improvement metrics adjusted to common basis by conversion methodology assuming individual component efficiencies for a reference plant as follows: 87% boiler efficiency, 40% turbine efficiency, 98% generator efficiency, and 6% auxiliary load. Based on these assumptions, the reference power plant has an overall efficiency of 32% and a net heat rate of 10,600 Btu/kWh. As a result, if a particular efficiency improvement method was reported to achieve a 1% point increase in boiler efficiency, it would be converted to a 0.37 % point increase in overall efficiency. Likewise, a reported 100 Btu/kWh decrease in net heat rate would be converted to a 0.30% point increase in overall efficiency.

Exhibit 3-2. Summary of NETL performance, cost, and CO_2 emissions comparison analysis for nominal 550 MWe PC-fired EGU burning bituminous coal by steam cycle

Parameter	Bituminous Coal-Fired EGU	
	Subcritical Boiler	Supercritical Boiler
Gross Power Output (kWe)	583,315	580,260
Auxiliary Power Requirement (kWe)	32,780	30,110
Net Power Output (kWe)	550,445	550,150
Coal Flow Rate (lb/hr)	437,699	411,282
HHV Thermal Input (kW) Net Plant	1,496,479	1,406,161
HHV Net Efficiency (%)	36.8%	39.1%
Total Plant Cost ($ x 1,000)[a]	$852,612	$866,391
Total Plant Cost ($/kW)[a]	$1,549	1,575
Levelized Cost of Electricity (mills/kWh)[b]	64.0	63.3
CO_2 Emissions (ton/hr)	519.1	487.6
CO_2 Emissions (ton/year)	3,864,884	3,632,301
CO_2 Emissions (lb/MMBtu)	203	203
CO_2 Emissions (lb/MWh gross output)	1,780	1,681
CO_2 Emissions (lb/MWh net output)	1,886	1,773

Source: NETL, 2007. Exhibit ES-2.

[a] The NETL costs are presented as "overnight costs" in January 2007 dollars and do not include escalation, owner's costs, taxes, site specific considerations, labor incentives, etc.

[b] 10 mills are equivalent to 1 U.S. cent.

Continuing research and advances in metallurgy have allowed the development of supercritical boilers capable of operating at increasingly higher temperatures and pressures, achieving increasingly higher efficiencies. Ultra-supercritical (USC) boilers designed to operate at steam conditions in excess of 4,500 psi can potentially operate at efficiencies approaching 50% (HHV). Steam temperature and pressure selection for boilers depends in part upon fuel corrosiveness, and research is focused on the development of new materials for boiler tubes and high alloy steels that minimize corrosion. There are potential concerns that temperatures above 1,100°F (590°C) while firing high-sulfur coal (such as Illinois No. 6) would result in an exponential increase of the material degradation of the highest temperature portions of the superheater and reheater due to coal ash corrosion. This could require pressure parts replacement outages every 10 to 15 years. The availability and reliability of materials required to support the elevated temperature environment for high sulfur or chlorine applications, although extensively demonstrated in the laboratory, has not been fully demonstrated commercially (NETL, 2007). Additional factors that could limit steam temperatures and pressures are the maximum values specified in the American Society of Mechanical Engineers (ASME) Boiler and Pressure Vessel Code, Section I: Power Boilers. A developer deviating from this code could have difficulty acquiring insurance, or be out of compliance with specific state code requirements.

The commercial use of ultra-supercritical technology has historically been, and continues to be, prevalent in countries outside the U.S., such as Denmark, Germany, and Japan. Ultra-supercritical boilers burning various coal ranks are being widely deployed throughout the world. Construction of the first modern ultra-supercritical EGU built in the U.S. began in 2008 at the Southwest Electric Power Company's John W. Turk, Jr. Power Station near Texarkana, AR. This 600 MWe PC-fired facility will burn PRB subbituminous coal and is scheduled to begin operation in late 2012. Examples of ultra-supercritical PC-fired EGUs often cited as representing the currently highest efficiency operating coal-fired EGUs in the world include:

- 384 MWe ultra-supercritical PC-fired EGU for the European Vattenfall company's Nordjylland Power Station Unit 3 located near Aalborg, Denmark (Vattenfall, 2006). This power station began operation in 1998 and is a combined heat and power (CHP) facility that generates electricity and produces heat for a district heating system. The unit burns imported bituminous coal and uses seawater for cooling. Reported overall unit efficiency is 47% LHV or 45.3% HHV (International Energy Agency [IEA] Clean Coal Centre, 2007).
- 965 MWe ultra-supercritical PC-fired EGU for the German RWE Power company's Niederaussem Power Station Unit K located near Cologne, Germany (RWE Power, 2004). This unit burns lignite with a 51 to 58% moisture content and started operation in 2002. Reported unit operating efficiency is 43.2% LHV or 37% HHV (IEA Clean Coal Centre, 2007).
- Two 600 MWe ultra-supercritical PC-fired EGUs for the J-POWER (Electric Power Development Co., Ltd.) Isogo Thermal Power Station New Unit 1 and New Unit 2 located in Yokohama, Japan (J-POWER, 2009). These units burn domestic and imported bituminous coal and

use seawater for cooling. New Unit 1 started operation in 2002 and has a reported unit operating efficiency of 42% LHV or 40.6% HHV (IEA Clean Coal Centre, 2007). New Unit 2 started commercial operation in July 2009. The New No. 2 Plant, has a higher efficiency due to the boosting the reheat steam temperature 18°F higher than the New Unit No. 1 to 1,148°F (J-POWER, 2009).

- 450 MWe ultra-supercritical PC-fired EGU for the Capital Power Corporation's Genesee Power Station Unit 3 located near Edmonton, Alberta, Canada (Peltier, 2005). This unit burns Alberta subbituminous coal. This unit began operation in 2005 and has a reported unit operating net efficiency or 41% LHV or 39.6% HHV (IEA Clean Coal Centre, 2007).

As part of NETL's performance and cost baseline analysis for electricity production (NETL, 2007), IGCC EGUs were analyzed for the General Electric Energy (GEE), ConocoPhillips (CoP), and Shell coal gasification processes. The IGCC cases have different gross and net power outputs than the PC-fired EGU cases because of the combustion turbine size constraint. The advanced F-class turbine used to model the IGCC cases comes in a standard size of 232 MWe when operated on syngas. Each IGCC case uses two combustion turbines for a combined gross output of 464 MWe. Additional electrical output is generated by steam turbines, with steam from the HRSGs extracting heat from the combustion turbine exhaust. Although the two combustion turbines provide 464 MWe gross output in all cases, the overall combined cycle gross output ranges from 742 to 770 MWe. The net outputs range from 623 to 640 MWe depending on the gasification process. A summary comparison of the results for the three gasification processes without CO_2 CCS is presented in Exhibit 3-3. Although the efficiency of the combined cycle block is approximately 50% efficient in converting the syngas to electricity, parasitic loads of the gasification process lower the net efficiency for the IGCC to 38.2 to 41.1% HHV.

3.3.2. Coal Drying

Low-rank coals (lignite and subbituminous) are often utilized because the low cost per unit of heat input relative to bituminous coal and the low sulfur content. However, a major disadvantage of low-rank coals is their high moisture content, typically 25 to 40%. When this coal is burned, considerable energy is required to vaporize and heat the moisture, thus raising the heat rate of the EGU and lowering its efficiency. As fuel moisture decreases, the heating value of the fuel increases so that less coal needs to be fired to produce the same amount of electric power. Drier coal is also easier to handle, convey, and pulverize – reducing the burden on the coal-handling system. In addition, an EGU boiler designed for dried coal is smaller and has lower capital costs than a comparable EGU designed to burn coal that has not been dried. The pre-combustion drying of low-rank coals can improve the overall efficiency and several advanced coal drying technologies are or nearly are commercial available.

Exhibit 3-3. Summary of NETL performance, cost, and CO_2 emissions comparison for an IGCC power plant by gasification process

Parameter	Gasification Process		
	GEE	CoP E-	Shell
Gross Power Output (kWe)	770,350	742,510	748,020
Auxiliary Power Requirement	130,100	119,140	112,170
Net Power Output (kWe)	640,250	623,370	635,850
Coal Flow Rate (lb/hr)	489,634	463,889	452,620
HHV Thermal Input (kW) Net	1,674,044	1,586,023	1,547,493
HHV Efficiency (%)	38.2%	39.3%	41.1%
Total Plant Cost ($ x 1,000)[a]	1,160,919	1,078,166	1,547,483
Total Plant Cost ($/kW)[a]	1,813	1,733	1,977
Levelized Cost of Electricity (mills/kWh)[b]	78.0	75.3	80.5
CO2 Emissions (ton/hr)	561.9	539.1	527.1
CO2 Emissions (ton/year)	3,937,728	3,777,815	3,693,990
CO2 Emissions (lb/MMBtu)	197	199	200
CO2 Emissions (lb/MWh gross	1,459	1,452	1,409
CO2 Emissions (lb/MWh net	1,755	1,730	1,658

Source: NETL, 2007. Exhibit ES-2.

[a] The NETL costs are presented as “overnight costs” in January 2007 dollars do not include escalation, owner's costs, taxes, site specific considerations, labor incentives, etc.

[b] 10 mills are equivalent to 1 U.S. cent.

Great River Energy developed a coal drying technology for low-rank coals in partnership with the U.S. DOE as part of the DOE’s Clean Coal Power Initiative (U.S. DOE, 2007). The technology has been successfully demonstrated on a PC-fired boiler burning lignite at the utility's Coal Creek Station in Underwood, ND. The technology is now commercially offered under the trade name DryFiningTM (U.S. DOE, 2010b). The DryFiningTM process passes warm cooling water from the steam turbine exhaust condenser through an air heater where ambient air is heated before being sent to a fluidized bed coal dryer. The dried coal leaving the fluidized bed is sent to a pulverizer and then to the boiler. Air leaving the fluidized bed is filtered before being vented to the atmosphere. In addition to using power plant waste heat to reduce moisture, DryFiningTM also segregates particles by density. This means a significant amount of higher density compounds containing sulfur and mercury can be sorted out and returned to the mine rather than utilized in the boiler. The end result is that more energy can be extracted from the coal while simultaneously reducing emissions of mercury, sulfur dioxide, and NOX. At the Coal Creek Station, the process increased the energy content of the lignite from 6,200 to 7,100 Btu/lb, thereby resulting in a decrease in the fuel input into the boilers by 4% and a corresponding decrease in CO_2 emissions gains in overall efficiency of 2 to 4% are reported for the process.

RWE Power in Germany is also developing a fluidized bed drying technology for lignite, called WTA (RWE Power, 2009). A fundamental difference between the two drying processes is the WTA process first mills then dries the lignite while the DryFining^{TM} process first dries then mills the lignite. A prototype commercial-scale drying plant using the WTA process began operation in 2009 at the utility's Nederaussem Power Station site. For the WTA process, lignite is first milled to a fine particle size by hammer mills in direct series with a two-stage fluidized-bed dryer. The dried fuel exiting the stationary bed is separated from the gas stream and mixed with coarser lignite solids collected from the bottom of the dryer bed and then fed directly to the boiler. The heat needed for the drying of the fuel is supplied by external steam, which is normally taken from the turbine with the heat transfer taking place in tube bundles located inside the bed. Based on the development work completed to date of the WTA technology, the net gain in cycle efficiency is reported to be on the order of 4 percentage points, depending on the moisture content of the raw coal and the final moisture of the dried lignite.

Several other coal drying technologies are in ongoing development. One coal drying process being developed by DBAGlobal Australia Pty, Ltd., with the trade name Drycol process uses the controlled application of microwave radiation to dry coal (Graham, 2007). Coal feed stock is first separated into fine grade coal and one or more larger grades. The fine coal is loaded onto a conveyor and conveyed continuously through a microwave-energized heating chamber for drying. The fine grade coal is dried sufficiently so that when it is recombined with the larger grade coals, the moisture content of the aggregate coal is within a target moisture content range. Other coal drying technologies for low-rank coals in various stages of development include: 1) attrition milling of coal followed by air drying to produce a low-moisture coal product, 2) compressing heated, coarse crushed coal to squeeze water out , and 3) heating wet coal under pressure to approximately 480 to 570°F (APP, 2008).

To date, it has not been economic to dry subbituminous coal at the mine prior to transport to an EGU. In addition, subbituminous coal that has been dried results in increased coal dust, can spontaneously combust, and will reabsorb moisture during transport and storage. However, the development of more efficient drying technologies such as dryers using flue gas recirculation and briquetting of the dried coal to avoid spontaneous combustion and moisture reabsorption can improve the economics of upgrading low-rank coals. Descriptions of upgrading low rank coals are available from the White Energy Company (http://www.white energyco.com) and Evergreen Energy (http://www.evgenergy.com/). From an overall GHG perspective, the increased EGU efficiency and decreased transportation GHG emissions would have to be compared to the energy required to dry and process the coal at the mine.

3.3.3. Boiler Feedwater Heating & Hot-Windbox

The high-pressure liquid water entering the steam generator is called feedwater. A feedwater heater is an EGU component used to pre-heat water delivered to the boiler section. Thermodynamic optimization of this cycle is important to overall EGU efficiency. In a conventional EGU, the energy used to heat the feedwater is steam extracted between the stages of the steam turbine (see Exhibit 2-3). Therefore, approximately a quarter of the steam that would be used to perform expansion work in the turbine (and generate power) is not utilized for that purpose. However, using other heat sources for the feedwater heater avoids the need to extract steam from the turbine allowing the steam to be used for electric power generation and increases the output of the steam cycle and potentially lowers GHG emissions. This alternate heat source can either be from an integrated solar thermal energy source or from a combustion turbine. Examples of solar thermal energy used to augment the steam cycle at combined cycle facilities include the Martin Next Generation Solar Center in Florida and the proposed Green Energy Partners/Stonewall, LLC facility in Virginia. The first coal-fired power plant to integrate solar thermal technology is the Cameo generating station in Colorado. In addition,

EPRI (Electric Power Research Institute) is currently evaluating adding solar thermal energy to the Escalante and Mayo coal-fired power plants. An example of combustion turbine integration for feedwater heating is the Kettle Falls Generating Station (Schimmoller, 2003). For coal-fired boiler systems optimized to accommodate the combustion turbine exhaust, the incremental fuel efficiencies would be expected be comparable with combined cycle generation (Escosa, 2009; Stenzel). Another potential approach to integrate the use of a combustion turbine with a coal-fired steam cycle is using the turbine exhaust directly in the boiler in a hot-windbox. This involves injection of the combustion turbine exhaust directly into the boiler windbox or primary air ducts to provide an oxygen source as well as a heat source.

3.4. Combined Heat and Power Plant

Coal-fired EGUs dedicated to electric power generation and using the latest commercially available advanced technologies will generally operate at overall net efficiencies of approximately 40%. Significant amounts of energy released by coal combustion are lost during the steam condensation segment of the Rankine cycle due to heat transfer into the cooling water. In Europe, electricity is commonly generated by facilities that serve as both electricity generators and thermal energy producers for the local town or city district heating system. These combined heat and power (CHP) facilities are also known as cogeneration facilities. Operating an electric power station in a CHP mode allows recovery of some of the heat that would otherwise be rejected into cooling water, improving the overall efficiency of energy utilization. In applying CHP to an existing or new EGU, the temperature of the cooling water is normally not high enough to meet the requirements for most district heating or industrial process applications. In these cases, steam would be extracted at an elevated pressure and temperature from an intermediate stage of the steam turbine and then used for district or process heating.

This results in a decrease in the total electric power generation from the EGU. However, the overall fuel efficiency of CHP is higher than if electricity and steam were generated separately.

Because electricity can be transmitted over long distances, electric power plants can be located in remote areas as well as urban areas. However, thermal energy cannot be effectively transported over extended distances. This limits the practicality of incorporating a CHP mode into many electric power plant designs. The EGU needs to be located in close proximity to either a district energy system or an industrial facility with a significant and steady thermal demand. There are a number of examples; however, where industrial facilities have collocated with existing or new coal power plants in order to have access to reliable, low cost steam:

- DuPont's titanium dioxide plant in Johnsonville, Tennessee, is located next to TVA's Johnsonville power plant and buys high pressure process steam from the 1,200 MW facility. The power plant is comprised of 10 coal-fired boiler steam turbine units; the DuPont plant uses steam extracted from Units 1 through 4. Providing steam to the DuPont facility at the required process pressures reduces overall output of the power plant by 50 MW.
- Blue Flint Ethanol in Underwood, North Dakota, is a 50 million gallon per year dry mill ethanol producer located next to Great River Energy's 1,160 MW Coal Creek lignite-fired power plant. Starting operations in 2007, the ethanol facility purchases approximately 100,000 pounds per hour of medium pressure steam extracted from the power plant.
- Goodland Energy Center in Goodland, Kansas, is a small 22 MW coal-fired power plant that is supplying steam to a 20 million gallon per year ethanol plant and a 12 million gallon per year biodiesel plant, both co-located with the power plant. The power plant and the ethanol plant both started construction in 2006.

3.5. Oxygen Combustion

Oxygen combustion (oxy-combustion, oxy-firing or oxy-fuel) is an emerging technology applicable to either new or existing EGUs. The advantage offered by this technology is its potential for CO_2 emissions control because it produces a concentrated (nearly pure) CO_2 exhaust gas stream that requires minimal post-combustion clean-up prior to compression, transportation, and injection for long term storage. The basic concept of oxy-combustion is to use a mixture of oxygen (or oxygen-enriched air) and recycled flue gas (containing mostly CO_2) in place of ambient air for coal combustion. The resulting flue gas contains primarily CO_2 and water vapor with smaller amounts of oxygen, nitrogen, SO_2, and NOX. Consequently, the flue gas can be processed relatively easily to further purify the CO_2 (if necessary) for use in enhanced oil or gas recovery or for geological storage.

An oxy-combustion power plant consists of an air separation unit (ASU), an EGU with O_2-blown combustion, and a CO_2 treatment unit. The conventional ASU is a cryogenic process that has a significant energy requirement. However, alternative oxygen separation methods are being researched for possible commercial scale development. These alternative methods include ion transport membranes (ITM), ceramic autothermal recovery, oxygen transport membranes, and chemical looping (UARG, 2008). Oxygen is mixed with recirculated flue gas to create a mixture of O_2 and CO_2 (and some H_2O) which is used as the source of combustion oxidant instead of ambient air. The absence of air nitrogen produces a flue gas stream with a high concentration of CO_2.

Several research institutes are focusing on laboratory- and pilot-scale testing of oxy-fuel combustion (Levasseur, 2009). Pilot test programs currently are being conducted for European Enhanced Capture of CO_2 (ENCAP) program and the Advanced Development of the Coal-Fired Oxyfuel Process with CO_2 Separation (ADECOS) program. Additional research and development programs are being conducted, including:

- A 30 MW oxy-firing pilot plant at the Schwarze Pumpe station in Spremberg, Germany. This plant is the first complete oxy-combustion unit that includes the integrated system from the air separation unit to the gas purification and compression systems. The CO_2 will be compressed and liquefied for storage experiments to be conducted.
- A 32 MW oxy-firing demonstration project in France retrofitting an existing boiler to natural gas oxy-combustion. The captured CO_2 will be transported through an approximately 19 mile long pipeline and stored in a depleted gas field in Lacq, South of France.
- A comprehensive test program using the 15 MW tangentially-fired Boiler Simulation Facility and 15 MW Industrial Scale Test Facility operated by Alstom Power, Inc., in Windsor, CT. Testing is being conducted to assess a broad range of oxy-combustion design options. Project partners include the U.S. DOE, the Illinois Clean Coal Institute, and 10 electric utility companies.

4. Coal-Fired EGU Technology Alternatives Analysis

There is no one best available coal-fired EGU technology universally applicable to all EGU projects. The coal-fired EGU technology alternatives most suitable for a given project must be evaluated on a site-specific basis. An evaluation for a new facility would include the use of carbon capture and storage and the most efficient technologies (e.g., ultra-supercritical steam conditions, IGCC, pressurized fluidized bed), double steam reheat, coal drying, FGD technology, and CHP.

4.1. Site-Specific Coal-Fired EGU Technology Alternatives Analysis Example

This section summarizes the results for analyses prepared in support of an air permit application for a new 830 MW supercritical PC-fired EGU to be built by the Consumers Energy Company at their Karn-Weadock Generating Station in Bay County, Michigan. The EGU is designed to burn PRB subbituminous coal, but can also mix bituminous coal based on supply or price variations. A supercritical PC boiler with steam turbine throttle pressure of 3,805 psia and superheat and single reheat temperatures of 1,100 °F was proposed by the Consumers Energy Company.

The air permit application was initially submitted to the State of Michigan Department of Natural Resources and Environment (DNRE) in 2007. As an amendment to the permit application for the Consumers Energy project, the DNRE requested that Consumers Energy Company prepare and submit a top-down BACT analysis for IGCC plant as an alternative to building the supercritical PC-fired EGU. The supercritical PC-fired EGU air emissions controls included selective catalytic reduction (SCR), baghouse, wet flue gas desulfurization, hydrated lime injection, and activated carbon injection. The IGCC plant air emissions controls included SCR, Selexol or Sulfinol, black water handling equipment, and sulfur-impregnated activated carbon.

To estimate the IGCC plant capital costs, the permit applicant developed a premium factor for constructing an IGCC plant compared to a similar capacity supercritical PC-fired EGU based on information collected by the permit applicant including reported costs of other IGCC projects. This capital cost premium factor was then applied to the estimated costs for the supercritical PC-fired EGU. Additional cost estimates were made for fuel, water consumption, waste disposal, operation and maintenance, and pollutant allowance purchases required for each facility type. The reference IGCC plant capital costs were estimated to be approximately 24% higher than the supercritical PC-fired EGU. The analysis estimated that the cost of electricity generation from an IGCC unit would be approximately 37% higher for the IGCC unit than for the supercritical PC-fired EGU. The projected cost of generation for the supercritical PC-fired EGU was $60/MWh compared to $95/MWh for the IGCC plant. The cost estimates prepared for the analysis are summarized in Exhibit 4-1.

At the request of the State air permitting authority for the project, the permit applicant prepared and submitted a second electric generation alternatives analysis for the Consumers Energy project in June 2009 (Consumers Energy Company, 2009). As part of this alternatives analysis, a comparison of the various coal-fired EGU technologies was presented with respect to air emissions from coal combustion including CO_2. The ultra-supercritical PC-fired EGU has the lowest projected heat rate of the analyzed technologies. Of the technologies without carbon capture and storage, it also has the lowest GHG emissions rate. In December 2009, the Michigan Department of Environmental Quality issued the construction permit based on the proposed supercritical PC boiler. The results are summarized in Exhibit 4-2.

Exhibit 4-1. Supercritical PC-fired EGU and IGCC plant cost comparison Summary prepared for Consumers Energy EGU project

Cost Parameter[a]	800 MWe net Supercritical PC-fired EGU	800 MWe net IGCC Plant	Difference (IGCC vs. PC)
Capital Costs ($)	$2,671,916,111	$3,526,039,934	$854,123,823
Annualized Costs ($/yr)	$273,871,401	$361,419,093	$87,547,692
Fuel Cost ($/yr)	$117,624,738	$130,679,946	$13,055,208
Cooling Water Consumption Cost ($/yr)	$5,166,656	$3,114,268	(-$2,052,388)
Waste Disposal Cost ($/yr)	$316,937	$201,042	(-$115,895)
Operating and Maintenance Costs ($/yr)	$59,307,369	$84,336,386	$25,029,017
Total Annual Cost ($/yr)	$456,287,101	$579,750,735	$123,463,634
Annualized Cost of SO_2 Allowances ($/yr)	$2,442,975	$610,531	(-$1,832,445)
Annualized Cost of NO_x Allowances ($/yr)	$4,132,417	$2,611,787	(-$1,520,630)

Source: Consumers Energy Company, 2008.

[a] Costs are presented in September 2007 dollars and include owner's and financing costs.

Exhibit 4-2. Coal-fired EGU technology alternatives cost comparison summary prepared for Consumers Energy EGU project

Coal-Fired EGU Technology	40-yr. BusBar Cost Excluding CO2 Cost[a]	40-yr. BusBar Cost Including CO2 Cost[a]	Availability	Efficiency-Heat Rate (Btu/kWh)	CO2 Emission Rate (tons/MWh)
Supercritical PC-fired 830 MW	$97 per MWh	$133 per MWh	86% to	9,134	0.94
Subcritical PC-fired	$101 per MWh	$136 per MWh	84% to	9,407	0.97
Subcritical CFB boiler	$108 per MWh	$145 per MWh	87%	9,798	1.01
Supercritical CFB boiler	$108 per MWh	$144 per MWh	87%	9,508	0.98
IGCC plant	$128 per MWh	$162 per MWh	70% to	9,490	0.93
Ultra-supercritical PC-fired	$98 per MWh	$133 per MWh	91%	9,019	0.93
Supercritical PC-fired with Carbon Capture and Storage	$135 per MWh	$139 per MWh	86% to 91%	10,836	0.11
Supercritical PC-fired 500 MW	$104 per MWh	$140 per MWh	86% to	9,134	

Source: Consumers Energy Company, 2009.

[a] Busbar costs is the cost to generate the power leaving the plant (beyond the generator but prior to the voltage transformation point in the plant switchyard) and include all plant fixed costs (including all costs associated with the capital investment), fuel costs, operating and maintenance costs, emissions costs, interconnection costs, and transmission system upgrade costs. A busbar cost excluding CO_2 costs assumes a CO_2 tax or cap-and-trade program has not been implemented. A busbar cost including CO_2 costs assumes a CO_2 tax cost of $22/ton beginning in 2012 and rising to $53/ton by 2025.

4.2. EPA GHG Mitigation Database

The EPA Office of Research & Development (ORD) is collecting information regarding CO_2 mitigation measures applicable to coal-fired EGUs for compilation in a publicly-accessible GHG Mitigation Database. Version 1 of this database is expected to be released to the public in late summer 2010. The database is a tool that provides information of both commercially available technologies, as well as emerging technologies that are being demonstrated at larger scales for commercial viability.

REFERENCES

Internet Web Site addresses cited for individual references were available as of the date of publication of this document.

Asai, Akihisu, et al. 2004. *System Outline and Operational Status of Karita Power Station New Unit 1 (PFBC)*. JSME International Journal, Series B. Vol. 47, No. 2, 2004. Pp. 193-199. Available at: <http://www.jstage.jst.go.jp/article/jsmeb/47/2/193/_pdf>.

Asia-Pacific Partnership on Clean Development and Climate (APP). 2008. *Brown Coal Drying Technologies*. APP Joint Meeting of Cleaner Fossil Energy and Power Generation and Transmission Task Forces. April 1, 2008. Available at: <http://www.asiapacificpartnership.org/pdf/CFE/meeting_melbourne/SurveyofBrownCoalDr yingTechnologies-Godfrey.pdf>.

Benson, Lewis, et al. 2003. *Control of Sulfur Dioxide and Sulfur Trioxide Using By-Product of a Magnesium-Enhanced Lime FGD System*. Presented at ICAC Forum '03 Multi-Pollutant Emission Controls & Strategies, Multi-Pollutant Emission Controls & Strategies, Nashville, TN. October 14-15, 2003. Available at: <http://www.carmeusena.com/files/files/techpapersreports/carmeuse_icac_20forum_2003.pd f>.

Benson, Lewis. *New Magnesium-Enhanced Lime Flue Gas Desulfurization Process*. Technical Paper Report, Carmeuse North America. Pittsburgh, PA. Available at: <http://www.carmeusena.com/files/files/TechPapers Reports/fgd_new_magnesium.pdf >.

Consumers Energy Company. 2008. *Amendment to Application No. 341-07 Consumers Energy Company – Bay Count Top-Down BACT Analysis for IGCC*. April 2008. Consumers Energy Company, Jackson, MI. Available at: <http://www.deq.state.mi.us/aps/downloads/permits/CFPP/2007/341-07/TopDown%20BACT%20Analysis%20for%20IGCC.pdf >.

Consumers Energy Company. 2009. *Balanced Energy Initiative: Electric Generation Alternatives Analysis*. June 2009. Consumers Energy Company, Jackson, MI. Available at: <http://www.deq.state.mi.us/aps/downloads/permits/PubNotice/341- 07/Alternatives Analysis.pdf >.

Davenport, W.G., et al. 2006. *Sulfuric Acid Manufacture*. Southern African Pyrometallurgy. South African Institute of Mining and Metallurgy, Johannesburg, 5-8 March 5-8, 2006.

Escosa, Jesús M. and Luis M. Romeo. 2009. *Optimizing CO_2 Avoided Cost by Means of Repowering*. Applied Energy, 86 (2009) 2351–2358.

Foster Wheeler North America Corp. 2009. *Utility CFB Goes Supercritical - Foster Wheeler's Lagisza 460 MWe Operating Experience and New 600 - 800 MWe Designs*. Prepared by James Utt, Arto Hotta, and Stephen Goidich, Foster Wheeler North America Corp., Clinton, NJ, for presentation at Coal-Gen 2009, Charlotte, NC, August 19-21, 2009. Available at: <http://www.fwc.com/publications/tech_papers/files/TP_CFB_09_12.pdf>.

Graham, James. 2007. *Microwaves for Coal Quality Improvement: The Drycol Project*. DBAGlobal Australia, Milton Queensland, Australia. Presented at the SACPS/International Pittsburgh Coal Conference 2007, Johannesburg, South Africa, September 10-14, 2007.

Available at: <http://www.drycol.com/downloads/Drycol%20Paper%20ACPS1%20060608. pdf>.

He, Boshu, et al. 2002. *Temperature Impact on SO_2 Removal Efficiency by Ammonia Gas Scrubbing*, Energy Conversion and Management 44:2175-2188. http://www.sciencedirect.com/science?_ob=ArticleURL&_udi=B6V02-3WRC6DR3&_ user=775537&_origUdi=B6V2P-47CY482-G&_fmt=high&_coverDate=07%2F31%2F1998&_rdoc=1&_orig=article&_acct=C0000429 38&_version=1&_urlVersion=0&_userid=775537&md5=e2354e5058cba8f383722059ffecd1 cd

Hong, B.D. and E.R. Slatick. 1994. *Carbon Dioxide Emission Factors for Coal*. Originally published in U.S. Energy Information Administration, Quarterly Coal Report, January-April 1994, DOE/EIA-0121(94/Q1). Washington, DC, August 1994, pp. 1-8. Available at: <http://www.eia.doe.gov/cneaf/coal/quarterly/co$_2$_article/co2.html>.

IEA Clean Coal Centre. 2007. *G8 Case Studies by the IEA Clean Coal Centre*. Prepared by Colin Henderson, IEA Clean Coal Centre, London, United

Kingdom for presentation at Third International Conference on Clean Coal Technologies for Our Future, Sotacarbo Coal Research Centre, Carbonia Sardinia, Italy. May 15 -17, 2007. Available at:<http://www.iea-coal.org/publishor/system/component_view.asp?LogDocId=81704&PhyDocId=6360>.

J-POWER (Electric Power Development Co., Ltd.). 2009. *Annual Report 2009*. Electric Power Development Co., Ltd., Tokyo, Japan. October 2009. Available at: <http://www.jpower.co.jp/english/ir/pdf/2009.pdf>.

Kaplan, P. Ozge, et al. 2008. *Is It Better To Burn or Bury Waste for Clean Electricity Generation?*. Environmental Science & Technology, Vol. 43, No. 6, 2009, pp. 1711-1717.

Korhonen, S., et al. 2001. Methane and Nitrous Oxide Emissions in the Finnish Energy Production. Fortum Tech-4615.

Levasseur, A. A., et al.. 2009. *Alstom's Oxy-Firing Technology Development and Demonstration - Near Term CO_2 Solutions*. Presented at the 34th International Technical Conference on Clean Coal & Fuel Systems, Clearwater, FL. May 31 - June 4, 2009. Available at: <http://www.netl.doe.gov/technologies/coalpower/ewr/co_2/pubs/5290%20Alstom%20oxycombustion%20paper%20Clearwater%20jun09.pdf>.

Maziuk, John and John Kumm. 2002. *Comparison of Dry Injection Acid-Gas Control Technologies*. Presented at the 95th Annual Conference and Exhibition of the Air and Waste Management Association, Baltimore, MD. June 23-27, 2002.

Michigan Department of Natural Resources and Environment. 2009. PTI Application No. 341-07. December 29, 2009. Available at: <http://www.deq.state.mi.us/aps/downloads/permits/CFPP/2007/341-07/341-07.htm>.

National Energy Technology Laboratory (NETL), 2007. *Cost and Performance Baseline for Fossil Energy Plants, Volume 1: Bituminous Coal and Natural Gas to Electricity, Revision 1*. DOE/NETL-2007/1281. U.S. Department of Energy, National Energy Technology Laboratory, Pittsburgh, PA. August 2007. Available at: <http://www.netl.doe.gov/energyanalyses/pubs/Bituminous%20Baseline_Final%20 Report.pdf>,

National Energy Technology Laboratory (NETL), 2008. *Reducing CO_2 Emissions by Improving the Efficiency of the Existing Coal-fired Power Plant Fleet*, DOE/NETL-2008/1329. U.S. Department of Energy, National Energy Technology Laboratory, Pittsburgh, PA. July 23, 2008. Available at: <http://www.netl.doe.gov/energy-analyses/pubs/CFPP%20Efficiency-FINAL.pdf>.

National Energy Technology Laboratory (NETL), 2010a. *Overview of DOE's Gasification Program*. Presentation by Jenny Tennant, Technology Manager, Gasification, U.S. U.S. Department of Energy, National Energy Technology Laboratory, Pittsburgh, PA. January 25, 2010. Available at: <http://www.netl.doe.gov/technologies/coalpower/gasification/pubs/pdf/DOE%20Gasification%20Program%20Overview%202010%2001-5%20v1v. pdf>.

National Energy Technology Laboratory (NETL), 2010b. *CCPI/Clean Coal Demonstrations Nucla CFB Demonstration Project, Project Fact Sheet*. U.S. Department of Energy, National Energy Technology Laboratory, Pittsburgh, PA. Accessed June 21, 2010. Available at: <http://www.netl.doe.gov/technologies/coalpower/cctc/summaries/nucla/nuclademo.html>.

National Energy Technology Laboratory (NETL), 2010c. *CCPI/Clean Coal Demonstrations Tidd PFBC Demonstration Project, Project Fact Sheet*. U.S. Department of Energy, National Energy Technology Laboratory, Pittsburgh, PA. Accessed June 21, 2010. Available at: <http://www.google.com/imgres?imgurl=http://www.netl.doe.gov/technologies/coalpower/cctc/summaries/tidd/images/tidd_plant_bw.jpg&imgrefurl=http://www.netl.doe.gov/technologies/coalpower/cctc/summaries/tidd/tidddemo.html&usg=__PMP6wAdGSJtmGwsCiNKKCwryUc=&h=>

Pacific Northwest National Laboratory (PNNL), 2009. *An Assessment of the Commercial Availability of Carbon Dioxide Capture and Storage Technologies as of June 2009*, PNNL18520. Pacific Northwest National Laboratory, Richland, WA. June 2009. http://www.pnl.gov/main/publications/external/technical_reports/PNNL-18520.pdf>.

Peltier, Robert. 2005. *Genesee Phase 3, Edmonton, Alberta, Canada*. Power. July/August 2005. Available at: <http://www.epcor.ca/SiteCollectionDocuments/Corporate/pdfs/publications%20and%20ne wsletters/plattspower05.pdf>.

Peltier, Robert. 2010. *Plant Efficiency: Begin with the Right Definitions*. Power. February 1, 2010. Available at: <http://www.powermag.com/gas/Plant-Efficiency-Begin-with-the-RightDefinitions_2432.html>.

Retzlaff, Klaus M and W. Anthony Ruegger, 1996. *Steam Turbines for Ultrasupercritical Power Plants*. GE Power Generation, GER-3945A. Available at: <http://www.gepower.com/prod serv/products/tech docs/en/downloads/ger 3945a.pdf>.

RWE Power. 2004. *Niederaussem Power Plant: A Plant Full Of Energy*. RWE Power Aktiengesellschaf, Essen, Germany. October 2004. Available at: <http://www.debriv.de/tools/download.php?filedata=1218532199.pdf&fil ename=Kraftwerk %20Niederaussem%20(englisch).pdf&mimetype=application/pdf>.

RWE Power. 2009. *WTA Technology: A Modern Process for Treating and Drying Lignite*. RWE Power Aktiengesellschaf, Essen, Germany. February 2009. Available at: <http://www.rwe.com/web/cms/mediablob/en/88166/data/183490/36906/rwe/innovations/power-generation/coal-innovation-centre/fluidized-bed-drying/download-wta-en.pdf>.

Sargent & Lundy. 2009. *Coal-Fired Power Plant Heat Rate Reductions*. Report Number SL-009597. Sargent & Lundy, Chicago, Il. January 22, 2009. Available at: <http://www.epa.gov/airmarkt/resource/docs/coalfired.pdf>.

Schimmoller, Brian, 2003. *Avista Kettle Falls*. Power Engineering. October, 2003. Available at: <http://www.powergenworldwide.com/index/display/articledisplay/189185/articles/powerengineering/volume-107/issue-10/departments/managing-the-plant/avista-kettle-falls.html>

Shand, Mark A., 2009. *A New Look at SO_2 Removal*. April 28, 2009. Available at: <http://www.risiinfo.com/technologyarchives/chemicalsnews>.

Srivastava, Ravi K, Wojciech Jozewicz and Carl Singer, 2001. *SO2 Scrubbing Technologies: A Review*. Environmental Progress, 20(4):219-28.

Srivastava, Ravi K and Wojciech Jozewicz. 2001. *Flue Gas Desulfurization: The State of the Art*, Journal of the Air & Waste Management Association, 51:1676-88.

Stenzel, William C., et al. *Repowering Existing Fossil Steam Plants*. December 11, 1997. Available at: <http://soapp.epri.com/press/default.htm>.

Utility Air Regulatory Group (UARG). 2008. *A Review of Carbon Capture and Sequestration (CCS) Technology*, Prepared by J. Edward Cichanowicz, Saratoga, CA, December 2008.

U.S. Department of Energy (DOE), 1996. *Electric Utility Engineer's FGD Manual, Vol. 1--FGD Process Design*. U.S. Department of Energy, Office of Fossil Energy, Morgantown, WV. March 1996. Available at: <http://www.netl.doe.gov/technologies/coalpower/ewr/pubs/fgdmanual_vol1.pdf>.

U.S. Department of Energy (DOE). 2007. *Clean Coal Technology Power Plant Optimization Demonstration Projects Topical Report Number 25*.

September 2007. Available at: <http://www.netl.doe.gov/technologies/coalpower/cctc/topicalreports/pdfs/topical25.pdf>

U.S. Department of Energy (DOE), 2009. *Technical Workshop Report: Opportunities to Improve the Efficiency of Existing Coal-Fired Power Plants.* Workshop sponsored by the U.S. Department of Energy and National Energy Technology Laboratory. Rosemont, IL.

July 15-16, 2009. Available at: <http://www.netl.doe.gov/energy-analyses/pubs/NETL%20Power%20Plant%20Efficiency%20Workshop%20Report%20Final.pdf>.

U.S. Department of Energy (DOE), 2010. *Technical Workshop Report: Improving the Efficiency of Coal-fired Power Plants in the United States.* Workshop sponsored by the U.S. Department of Energy and National Energy Technology Laboratory. Baltimore, MD. February 24-25, 2010. Available at: <http://www.netl.doe.gov/energyanalyses/refshelf/detail.asp?pubID=306>.

U.S. Department of Energy (DOE). 2010b. *Innovative Drying Technology Extracts More Energy from High Moisture Coal.* Fossil Energy Techline. March 11, 2010. Available at: <http://fossil.energy.gov/news/techlines/2010/10006-CCPI Technology Goes Commercial.html>.

U.S. Energy Information Administration (U.S. EIA). 2008. *EIA-923 (Schedule 2) - Monthly Utility and Nonutility Fuel Receipts and Fuel Quality Data (2008 Final Data).* U.S. Energy Information Administration, Office of Coal, Nuclear, Electric and Alternate Fuels, U.S. Department of Energy, Washington, DC 20585. Accessed June 21, 2010. Available at: <http://www.eia.doe.gov/cneaf/electricity/page/eia423.html>.

U.S. Energy Information Administration (U.S. EIA). 2010. *Electric Power Annual 2008.* DOE/EIA-0348(2008). U.S. Energy Information Administration, Office of Coal, Nuclear, Electric and Alternate Fuels, U.S. Department of Energy, Washington, DC 20585. January, 2010. Available at: <http://www.eia.doe.gov/cneaf/electricity/epa/epa_sum.html>.

U.S. Environmental Protection Agency (EPA). 2001. *Database of information collected in the Electric Utility Steam Generating Unit Mercury Emissions Information Collection Effort.* OMB Control No. 2060-0396. U.S. Environmental Protection Agency, Office of Air Quality Planning and Standards, Research Triangle Park, NC. April 2001.

U.S. Environmental Protection Agency (EPA). 2005. *Multipollutant Emission Control Technology Options for Coal-fired Power Plants.* EPA. EPA-600/R-05/034. U.S. Environmental Protection Agency, National Risk

Management Research Laboratory. Research Triangle Park, NC. March 2005. Available at: <http://www.epa.gov/airmarkt/resource/docs/multipreport2005.pdf>.

U.S. Environmental Protection Agency (EPA). 2006. *Environmental Footprints and Costs of Coat-Based Integrated Gasification Combined Cycle and Pulverized Coal Technologies*. EPA. EPA-600/R-06/006. U.S. Environmental Protection Agency, National Risk Management Research Laboratory. Research Triangle Park, NC. July 2006. Available at: <http://www.epa.gov/oar/caaac/coaltech/2007_01_epaigcc.pdf>.

Vattenfall. 2006. *Nordjylland Power Station: The World's Most Efficient Coal-Fired CHP Plant*. Vattenfall, A/S Copenhagen, Denmark. Available at: <http://www.vattenfall.dk/da/file/nordjyllanduk080121_7841588.pdf>.

VGB Powertech and Evonik Energy Services (2008). Supercritical and ultra supercritical technology. In Power Plant Performance Reporting and Improvement under the Provision of the Indian Energy Conservation Act: Output 1.1: Best practice performance monitoring, analysis of performance procedures, software and analytical tools, measuring instrumentation, guidelines or best practice manuals and newest trends. (Annexure XIV). Indo German Energy Programme. Retrieved from <http://www.emtindia.net/PowerPlantComponent/Output1.1/Output1.1.pdf>.

Weijuan, Yang, et al. 2007. Nitrous Oxide Formation and Emission in Selective Non-Catalytic Reduction. Energy Power Engineering China, 228-232.

In: Emissions Control in Electricity Generation ISBN: 978-1-62257-830-6
Editors: J. L. Walker, A. Hernandez

Chapter 3

AIR EMISSIONS AND ELECTRICITY GENERATION AT U.S. POWER PLANTS*

United States Government Accountability Office

The Honorable Sheldon Whitehouse
Chairman
Subcommittee on Oversight
Committee on Environment and
Public Works United States Senate

Dear Mr. Chairman:

The United States depends on a variety of fuels to generate electricity, including fossil fuels (coal, natural gas, and oil), nuclear power, and renewable sources. Power plants that burn fossil fuels provide about 70 percent of U.S. electricity, but they also produce substantial amounts of harmful air emissions.[1] In particular, electricity generating units at fossil fuel power plants are among the largest emitters of sulfur dioxide and nitrogen oxides, which have been linked to respiratory illnesses and acid rain, as well as of carbon dioxide, the primary greenhouse gas contributing to climate change.[2] Of the three fossil fuels, coal is the most widely used fuel in the United States, providing about 45 percent of electricity in 2010, followed by natural gas,

* This is an edited, reformatted and augmented version of United States Government Accountability Office, Publication No. GAO-12-545R, dated April 18, 2012.

which provided about 24 percent.[3] Coal plays a critical role in the reliability of the electricity grid, especially in certain geographic areas, but coal-fired units also generally emit more air pollution than units burning natural gas or oil.

Under the Clean Air Act, the Environmental Protection Agency (EPA) establishes national ambient air quality standards for six pollutants that states are primarily responsible for attaining.[4] States attain these standards, in part, by regulating emissions of these pollutants from certain stationary sources, such as electricity generating units. Numerous Clean Air Act requirements apply to electricity generating units, including New Source Review (NSR), a permitting process established in 1977.[5] Under NSR, owners of generating units must obtain a preconstruction permit that establishes emission limits and requires the use of certain pollution control technologies. NSR applies to (1) generating units built after August 7, 1977, and (2) to existing generating units—regardless of the date built—that seek to undertake a "major modification," a physical or operational change that would result in a significant net increase in emissions of a regulated pollutant. Units built before August 7, 1977, are not required to undergo NSR unless they undertake a major modification. For the purposes of this report, we refer to units that began operation in or before 1978—the first full year after NSR was established—as "older units" and those that began operating after 1978 as "newer units."[6]

In limiting NSR's requirements to facilities built or undertaking major modifications after August 7, 1977, Congress allowed existing facilities to defer installation of pollution controls until they made a major modification, with the expectation that over time all facilities would either install such equipment or shut down, thereby lowering overall emissions. According to EPA data, 1,485 older units (43 percent of fossil fuel units) were still in operation in 2010. Some research suggests that many of these older units continue to operate without emissions controls, and in June 2002, we reported that older fossil fuel electricity generating units emitted air pollution at higher rates than newer units.[7]

This report responds in part to your request for information on electricity generation and emissions at U.S. electricity generating units and the implementation of NSR. Our objective is to provide information on how older fossil fuel electricity generating units compare with newer units in terms of their air emissions and electricity generation. To respond to this objective, we reviewed selected data elements in the Ventyx Velocity Suite EV Market-Ops database. This proprietary database contains consolidated energy and emissions data from EPA, the Energy Information Administration (EIA), and other sources. Specifically, we analyzed how older plants compare with newer plants in their emissions, energy production, location, and fuel type. We reviewed energy and emissions data from calendar year 2010. We reviewed data from units that (1) listed a fossil fuel (coal, natural gas, or oil) as a primary fuel; (2) generated electricity in 2010; and (3) had a net summer capacity greater than 25 megawatts, making them subject to EPA emissions monitoring and reporting requirements. In all, we examined the characteristics of 3,443 electricity generating units—1,485 older units and 1,958 newer units. We focused our analysis on power plant emissions of three regulated pollutants: sulfur dioxide (as a proxy for sulfur oxides), nitrogen oxides, and carbon dioxide (a greenhouse gas).[8] To assess the reliability of the Ventyx data, we reviewed existing documentation about the data and the system that produced them, interviewed Ventyx staff who were knowledgeable about the data, and consulted with EPA and EIA agency officials knowledgeable in energy issues. We determined the Ventyx data to be sufficiently reliable for the purpose of this report.

We conducted this performance audit from February 2012 to April 2012 in accordance with generally accepted government auditing standards. Those standards require that we plan and perform the audit to obtain sufficient, appropriate evidence to provide a reasonable basis for our findings and conclusions based on our audit objectives. We believe that the evidence obtained provides a reasonable basis for our findings and conclusions based on our audit objectives.

RESULTS IN BRIEF

Older electricity generating units—those that began operating in or before 1978—provided 45 percent of electricity from fossil fuel units in 2010 but produced a disproportionate share of emissions, both in aggregate and per unit of electricity generated. Overall, in 2010 older units contributed 75 percent of sulfur dioxide emissions, 64 percent of nitrogen oxides emissions, and 54 percent of carbon dioxide emissions from fossil fuel units. For each unit of electricity generated, older units collectively emitted about 3.6 times as much sulfur dioxide, 2.1 times as much nitrogen oxides, and 1.3 times as much carbon dioxide as newer units. The difference in emissions between older units and their newer counterparts may be attributed to a number of factors. First, 93 percent of the electricity produced by older fossil fuel units in 2010 was generated by coal-fired units. Compared with natural gas units, coal-fired units produced over 90 times as much sulfur dioxide, twice as much carbon dioxide and over five times as much nitrogen oxides per unit of electricity, largely because coal contains more sulfur and carbon than natural gas. Second, fewer older units have installed emissions controls, which reduce emissions by limiting their formation or capturing them after they are formed. Among coal-fired units—which produce nearly all sulfur dioxide emissions from electric power generation— approximately 26 percent of older units used controls for sulfur dioxide, compared with 63 percent of newer units. Controls for nitrogen oxide emissions were more common among all types of fossil fuel units, but these controls vary widely in their effectiveness. Among older units, 14 percent had installed selective catalytic reduction (SCR) equipment, the type of control capable of reducing the greatest amount of nitrogen oxides emissions, compared with 33 percent of newer units. In addition, approximately 38 percent of older units did not have any controls for nitrogen oxides, compared with 6 percent of newer units. Third, lower emissions among newer units may be attributable in part to improvements in the efficiency with which newer units convert fuel into electricity. Nonetheless, older units remain an important part of the electricity generating sector, particularly in certain regions of the United States.

BACKGROUND

The efficient and reliable operation of the electricity industry is critical to the health of the U.S. economy. Residential consumers rely on electricity to power their households, and electricity is a key input for businesses that produce trillions of dollars in products and services. Domestic energy production also provides U.S. jobs and supports economic growth. Plentiful, reliable, and relatively inexpensive fossil fuels have been a mainstay of the U.S. electricity mix.

The electricity generating sector is also a major contributor to air emissions in the United States. Based on emissions data published by EPA, in 2010, fuel combustion by electric utilities was responsible for about 65 percent of sulfur dioxide emissions and 16 percent of nitrogen oxides emissions from all sources nationwide.[9] Electricity generation was also responsible for 39 percent of carbon dioxide emissions from all sources in 2009.[10] Units in the Southeast, South Central, and Great Lakes regions produced most of the electricity and emissions from fossil fuel units. (See encls. I through IV for additional information on the location of generating units, their electricity production, and their emissions.)

As we have previously reported, sulfur dioxide and nitrogen oxides have been linked to a variety of health and environmental concerns, and carbon dioxide has been linked to climate change.[11] For example, sulfur dioxide and nitrogen oxides contribute to the formation of fine particles, and nitrogen oxides contribute to the formation of ozone.[12] Fine particles have been linked to premature death, aggravated asthma, and chronic bronchitis, and ozone can inflame lung tissue and increase susceptibility to bronchitis and pneumonia. In addition to affecting health, sulfur dioxide and nitrogen oxides reduce visibility and contribute to acid rain, which harms aquatic life and degrades forests. These emissions affect local air quality, and they can also travel hundreds of miles to affect the air quality of downwind states. In addition, the accumulation of carbon dioxide in the atmosphere is linked to increases in air and ocean temperatures, which could threaten coastal areas with rising sea levels, alter agricultural productivity, and increase the intensity and frequency of floods and tropical storms.

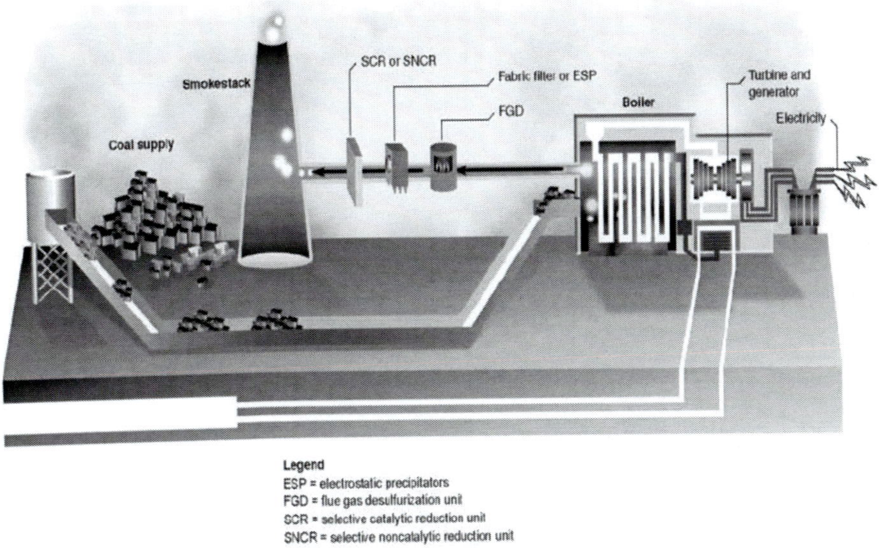

Source: GAO analysis information from Electric Power Research Institute and Tennessee Valley Authority.

Figure 1. Sample Layout of Emissions Controls at a Coal Power Plant.

Table 1. Summary of Emissions Control Equipment Used at Electricity Generating Units

Pollutant	Control equipment	How it works	Removal efficiency
Sulfur dioxidea	Flue gas desulfurization unit (commonly referred to as a "scrubber")	Wet flue gas desulfurization units inject a liquid sorbent, such as limestone, into the flue gas to form a wet solid that can be disposed of or sold. Dry flue gas desulfurization units inject a dry sorbent, such as lime, into the flue gas to form a solid by-product that is collected and removed.	Wet flue gas desulfurization units can remove 80-99 percent of sulfur dioxide. Dry flue gas desulfurization units can remove 70-95 percent of sulfur dioxide.
Nitrogen oxides	Combustion control technologies, such as low-nitrogen-oxides burnersb Postcombustion controls, such as selective catalytic reduction and selective noncatalytic reduction units	Coal combustion conditions are adjusted to inhibit the formation of nitrogen oxides. Selective catalytic reduction units inject ammonia into flue gas to form nitrogen and water and use a catalyst to enhance the reaction. Selective noncatalytic reduction units also inject ammonia but do not use a catalyst.	These technologies can reduce formation of nitrogen oxides by 40-45 percent. Selective catalytic reduction units can remove 70-95 percent of nitrogen oxides. Selective noncatalytic reduction units can remove 30-75 percent of nitrogen oxides.

Source: GAO.

[a]Another approach to reducing sulfur dioxide emissions from a coal-fired electricity generating unit is for a plant to switch from using coal with a higher sulfur content to coal with a lower sulfur content or to blend higher-sulfur coal with lower-sulfur coal.

[b]Low-nitrogen-oxides burners can be used in conjunction with postcombustion controls for nitrogen oxides.

Emissions controls can help reduce the emissions from generating units by either limiting their formation or capturing them after they are formed.[13] At coal-fired units, emissions controls are generally installed in a boiler, where coal is burned, or the duct work that connects the boiler to a smokestack. Figure 1 shows some of the pollution controls that may be used at coal power plants, including flue gas desulfurization units—known as scrubbers—to control sulfur dioxide emissions, fabric filters or electrostatic precipitators to control particulate matter, and selective catalytic reduction (SCR) or selective noncatalytic reduction (SNCR) units to control nitrogen oxides emissions. According to a 2010 EPA report, the development of effective and commercially viable carbon dioxide controls for coal-fired electricity generating units has received significant attention, but some of these technologies are still in the research and development phase, and most are not yet commercially viable.[14]

The reduction in emissions from the use of emissions controls can be substantial, as shown in table 1. Used commercially since the early 1970s, scrubbers are the most common technology for reducing sulfur dioxide emissions, and are capable of removing up to 99 percent of sulfur dioxide emissions. Commercially available low-nitrogen-oxide burners and SCRs are among the more common nitrogen oxide control technologies, with SCRs able to reduce these emissions by more than 90 percent. The installation of emissions controls can also be expensive. According to EPA, a typical coal unit with a capacity of 700 megawatts could incur from \$269 million to \$329 million to install a scrubber and from \$108 million to \$129 million to install an SCR, plus operating and maintenance costs.[15] Additionally, emissions controls can require additional electricity from the unit to operate, which reduces the amount of electricity available to be used by customers.

Older Fossil Fuel Units Produce More Emissions and Less Electricity Than Newer Units

In general, older electricity generating units, which run primarily on coal, produce more emissions than newer units, and at higher rates per unit of electricity. Much of this difference can likely be attributed to a greater reliance on coal among older units. In addition, fewer older units have installed emissions controls, and older units are generally less efficient at converting fuel to electricity. Nonetheless, some older units play an important role in providing reliable electricity to certain U.S. regions.

Older Units Emit More Than Newer Units

In 2010, older fossil fuel electricity generating units—those that began operating in or by 1978— produced a disproportionate share of emissions for the electricity they produced. Specifically, older units generated 45 percent of the electricity from fossil fuel units in 2010 but emitted 75 percent of sulfur dioxide emissions, 64 percent of nitrogen oxides emissions, and 54 percent of carbon dioxide emissions (see fig. 2). Units that began operating after 1978 were responsible for the remainder of the emissions and electricity production.[16]

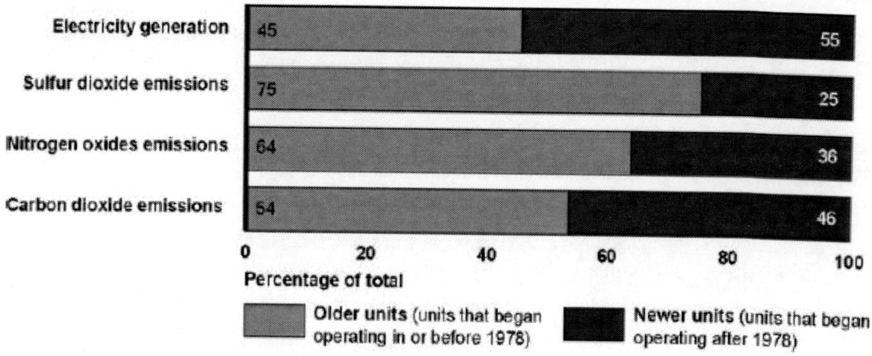

Note: Chart reflects older and newer generating units that use coal, natural gas, and oil. Older generating units are mostly coal fired, whereas newer generating units rely more heavily on natural gas.

Figure 2. Share of Total Electricity and Emissions per Unit of Electricity Generated by Fossil Generating Units in 2010.

Older units produced emissions at a higher rate than newer units in 2010; that is, they produced more emissions per unit of electricity they generated. As shown in figure 3, for each megawatt-hour of electricity generated, older units, on average, emitted about 3.6 times as much sulfur dioxide as newer units, about 2.1 times as much nitrogen oxides, and about 1.3 times as much carbon dioxide.

Emissions from fossil fuel units in 2010 varied widely, depending on the decade in which the unit began operating. In aggregate, the oldest units produced the most emissions, and more recently built units produced the least emissions (see fig. 4). For example, units that began operating in the 1960s generated 12 percent of electricity in 2010, but also produced 26 percent of sulfur dioxide emissions, 21 percent of nitrogen oxides emissions, and 15 percent of carbon dioxide emissions in that year. In comparison, units that began operating in the 2000s produced over twice as much electricity (26 percent) in 2010, but a much lower share of emissions— 1 percent of sulfur dioxide, 3 percent of nitrogen oxides, and 15 percent of carbon dioxide.

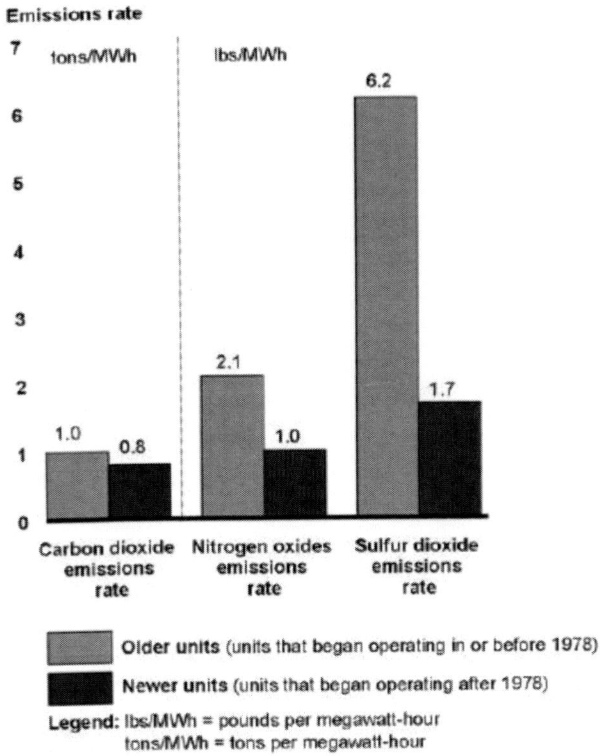

Source: GAO analysios of Ventyx data.

Note: Chart reflects older and newer generating units that use coal, natural gas, and oil. Older generating units are mostly coal-fired, whereas newer generating units rely more heavily on natural gas.

Figure 3. Emissions per Unit of Electricity Generated in 2010.

As shown in figure 5, units that began operating more recently are more likely to have comparatively lower emissions rates. In 2010, units that began operating in the 1970s or later generally emitted substantially less sulfur dioxide and nitrogen oxides for each unit of electricity they generated. For example, units built in the 1960s emitted an average of of 7.8 pounds of sulfur dioxide and 2.5 pounds of nitrogen oxides per megawatt-hour of electricity generated, whereas units built in the 2000s emitted an average of 0.1 pounds of sulfur dioxide and 0.2 pounds of nitrogen oxides per megawatt-hour.

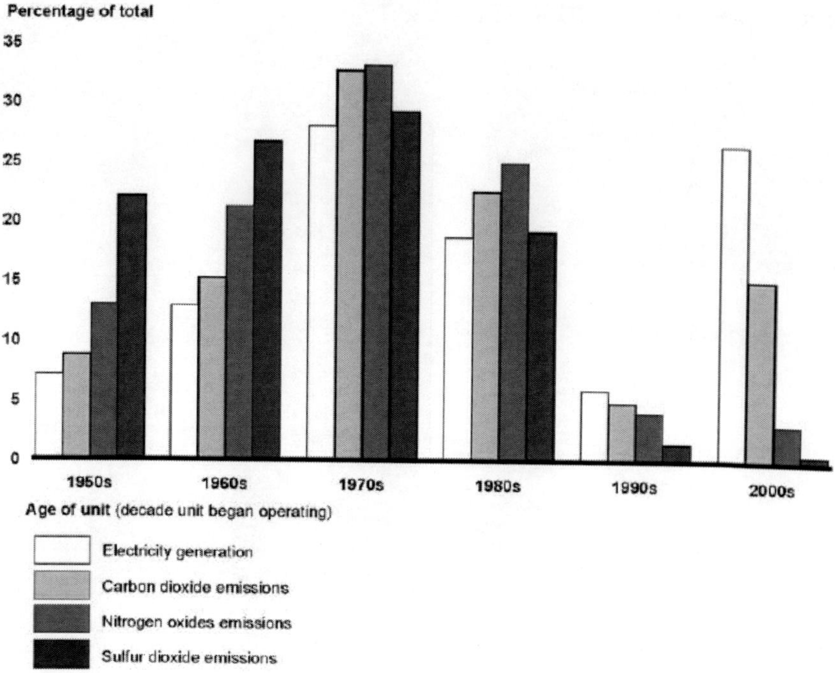

Source: GAO analysios of Ventyx data.

Note: Chart reflects generating units that use coal, natural gas, and oil. Older generating units are mostly coal-fired, whereas newer generating units rely more heavily on natural gas.

Figure 4. Share of Total Electricity and Emissions Produced by Fossil Fuel Generating Units in 2010, by Decade Unit Began Operating.

Carbon dioxide emissions rates have also declined among newer units in the last several decades, although to a smaller extent. As shown in figure 6, units that began operating in the 1980s emitted an average of 2,160 pounds of carbon dioxide per unit of electricity in 2010, a decline of around one percent compared to units that began operating in the 1950s. In contrast, units that began operating after the 1980s had substantially lower carbon dioxide emissions rates, on average. For example, units that began operating in the 2000s emitted carbon dioxide at a rate of 1,016 pounds per megawatt-hour, about 53 percent less than units that began operating in the 1980s.

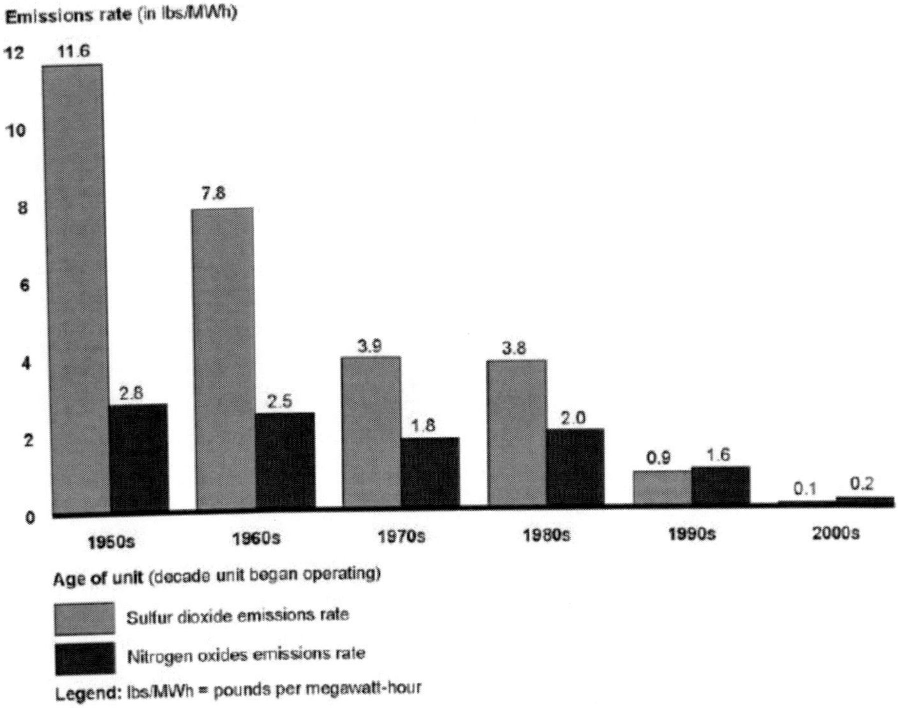

Source: GAO analysios of Ventyx data.

Note: Chart reflects generating units that use coal, natural gas, and oil. Older generating units are mostly coal-fired, whereas newer generating units rely more heavily on natural gas.

Figure 5. Emissions of Sulfur Dioxide and Nitrogen Oxides per Unit of Electricity Generated by Fossil Fuel Generating Units in 2010, by Decade Unit Began Operating.

As discussed below, we identified three reasons why older units may emit more than newer units: (1) older units are more likely to use coal, (2) older units are less likely to have installed emissions controls, and (3) older units are generally less efficient.

Older Units Are Mostly Fueled by Coal, and Newer Units Rely More on Natural Gas

The difference in emissions between older and newer units is likely due, in part, to significant changes in the fuels used to generate electricity—in particular, a shift from coal to natural gas.

Compared with coal, natural gas produces substantially lower emissions per unit of electricity generated, largely because natural gas contains less sulfur and carbon. This disparity is most apparent with sulfur dioxide emissions, as shown in figure 7.

On average, coal-fired units—both older and newer—produced over 90 times as much sulfur dioxide emissions per unit of electricity as natural gas-fired units in 2010.

Compared to natural gas-fired units, coal-fired units also produced over twice as much carbon dioxide and over five times as much nitrogen oxides per unit of electricity in 2010.

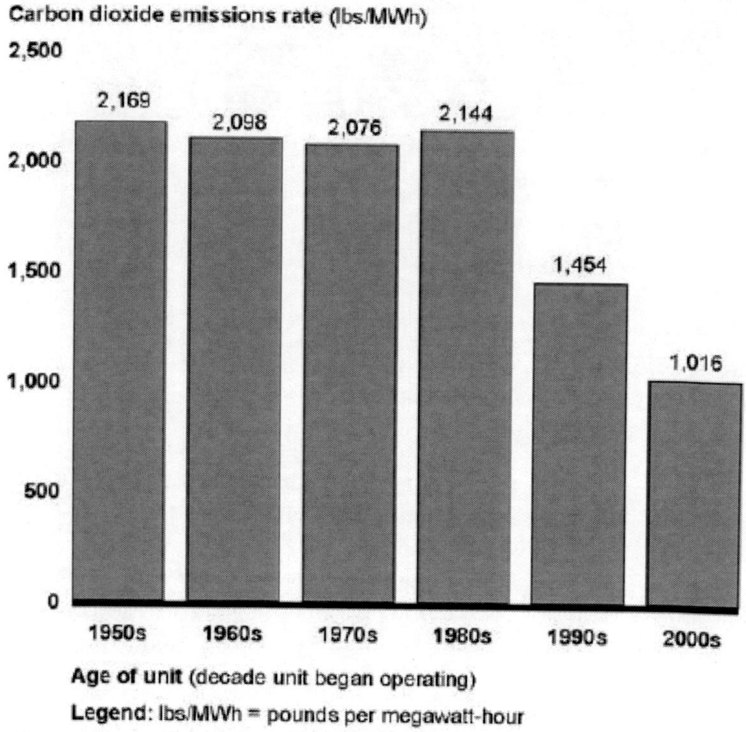

Source: GAO analysios of Ventyx data.

Note: Chart reflects generating units that use coal, natural gas, and oil. Older generating units are mostly coal-fired, whereas newer generating units rely more heavily on natural gas.

Figure 6. Carbon Dioxide Emissions per Unit of Electricity Generated by Fossil Fuel Generating Units in 2010, by Decade Unit Began Operating.

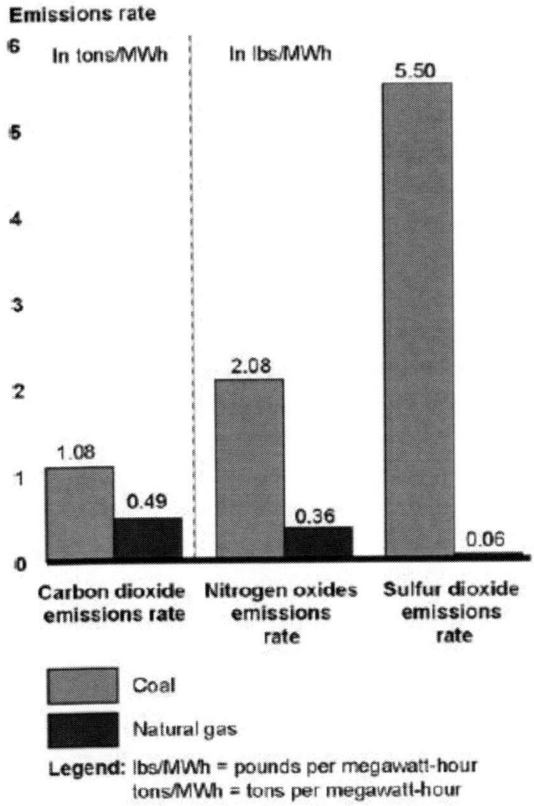

Source: GAO analysios of Ventyx data.

Figure 7. Emissions per Unit of Electricity Generated by Fossil Fuel Generating Units in 2010, by Type of Fuel.

Older generating units are much more likely than newer units to use coal as a primary fuel. In 2010, coal-fired units accounted for about 93 percent of electricity from older units and 47 percent of electricity from newer units. While other fossil fuels, particularly oil, played an important role in electricity generation in the late 1960s and 1970s, many of these units have since been retired, a development likely attributable to rising oil prices.[17]In addition, between 1978 and 1987, the Powerplant and Industrial Fuel Use Act of 1978 (Fuel Use Act) restricted construction of new power plants using oil or natural gas as a primary energy source.

Since the 1990s, natural gas has played a much larger role in electricity generation (see fig. 8). After the 1987 repeal of provisions in the Fuel Use Act, use of natural gas by the electric power sector increased by more than 240 percent.[18] As we have previously reported, factors contributing to the rise of natural gas include the comparatively low levels of emissions produced by natural gas units and a long period of low natural gas prices in the 1990s.[19] After rising in the early 2000s, natural gas prices have again fallen in recent years, a trend that may influence decisions about the future composition of the fleet of generating units, with the potential for retirements of older coal-fired units and the construction of new natural gas units.

In 2010, natural gas units accounted for approximately one-third of the electricity generated by fossil fuel units. Most of this natural-gas-based generation—about 78 percent—came from units that began operating in 2000 or later. Conversely, most coal-based generation—about 66 percent—came from units that have been operating more than 30 years.

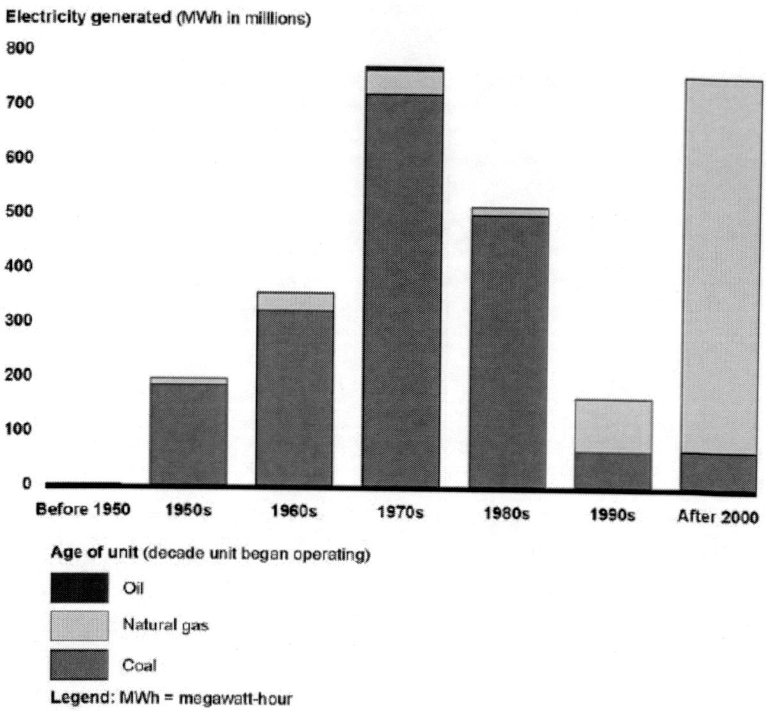

Source: GAO analysios of Ventyx data.

Figure 8. Electricity Generation from Fossil Fuel Generating Units in 2010, by Decade Unit Began Operating and Fuel Type.

Nonetheless, coal-fired units, both older and newer, remain both a key source of electricity and a significant source of emissions. As shown in figure

9, coal-fired units generated 67 percent of electricity from fossil fuel units in 2010, but also produced 99 percent of sulfur dioxide emissions, 92 percent of nitrogen oxides emissions, and 82 percent of carbon dioxide emissions from fossil fuel units in that year.

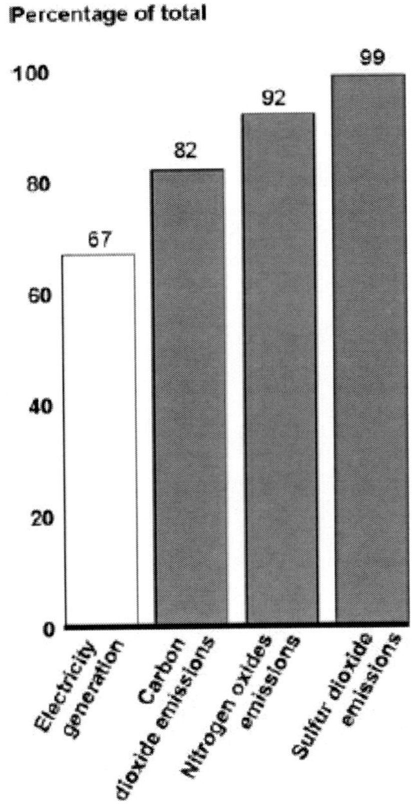

Source: GAO analysios of Ventyx data.

Figure 9. Electricity Generation and Emissions from Coal-Fired Units in 2010, as a Percentage of Total Generation and Emissions from Fossil Fuel Generating Units.

Older Units Are Less Likely to Use Emissions Controls

Another likely reason that older units have higher aggregate emissions is that many older units have not installed emissions controls. Over the last several decades, the electricity generating industry has employed a number of technologies to comply with NSR and other state and federal air quality regulations. Some of the most common and effective technologies include flue gas desulfurization unit systems, or "scrubbers," to control sulfur dioxide and SCRs to control nitrogen oxides.

Regarding the use of controls for sulfur dioxide in 2010, 63 percent of newer coal-fired units— those that began operating after 1978—had installed controls (see fig. 10). In contrast, 26 percent of older, coal-fired units had installed sulfur dioxide controls.[20]

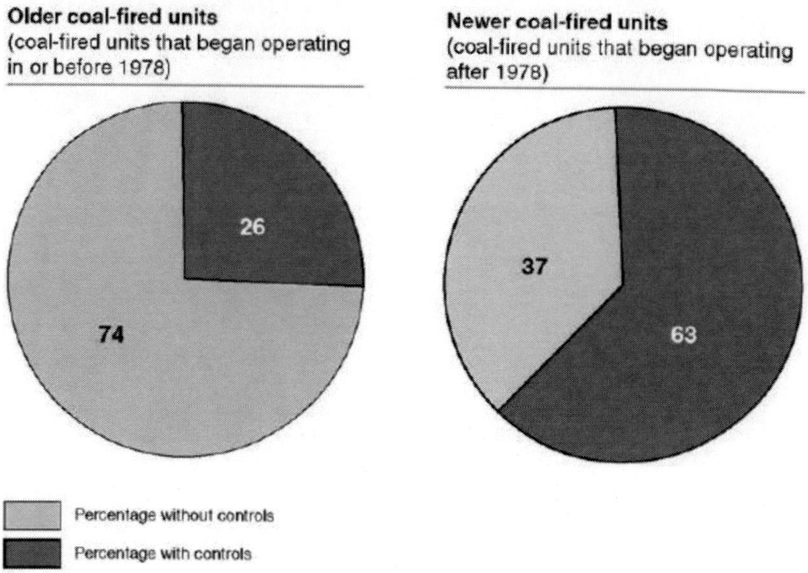

Source: GAO analysios of Ventyx data.

Figure 10. Use of Sulfur Dioxide Emissions Controls among Older and Newer Coal-Fired Generating Units in 2010.

Older coal-fired units without sulfur dioxide controls were responsible for the majority of sulfur dioxide emissions from fossil fuel units in 2010. Specifically, older coal-fired units without sulfur dioxide control equipment generated about 18 percent of electricity while producing 55 percent of sulfur dioxide emissions. As shown in figure 11, older coal-fired units without sulfur dioxide controls accounted for 2.8 million tons of sulfur dioxide emissions, nearly three times as much as older units with sulfur dioxide controls. Among those coal-fired units without sulfur dioxide controls, the majority (88 percent) were older units.

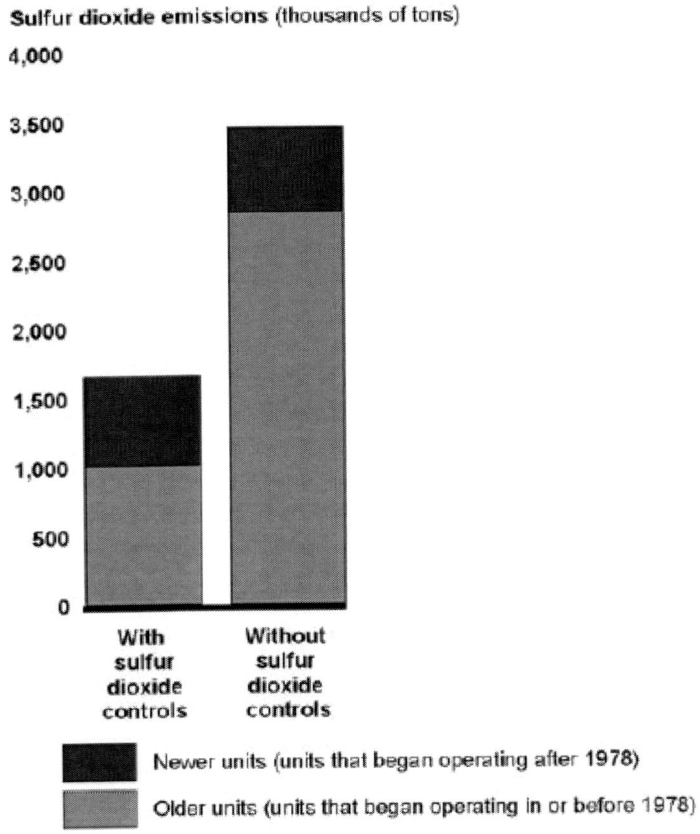

Source: GAO analysios of Ventyx data.

Figure 11. Sulfur Dioxide Emissions from Controlled and Uncontrolled Coal-Fired Generating Units in 2010.

The use of controls for nitrogen oxides is more widespread than for sulfur dioxide, although the effectiveness of these controls at removing nitrogen oxides varies widely. For purposes of our analysis, we classified controls for nitrogen oxides into three categories, as shown in table 2. Overall, older units were less likely than newer units to use controls for nitrogen oxides. Among older units, 38 percent had no controls for nitrogen oxides, compared with 6 percent of newer units. Older units were also less likely to have installed SCR controls, the most effective category of control for nitrogen oxides: about 14 percent of older units had such controls, compared with 33 percent of newer units.

Table 2. Controls for Nitrogen Oxides in Fossil Fuel Electricity Generating Units: Removal Efficiency and Extent of Use in 2010

Category	Control type	Number of older units	Number of newer units
1	None	564	116
2	Selective noncatalytic reduction, low nitrogen-oxides burners, and others	708	1,204
3	Selective catalytic reduction (SCR)	213	638

Source: GAO analysis of Ventyx data.

Note: Units in categories 2 and 3 may have installed multiple nitrogen oxides controls; no units in category 2, however, had installed an SCR.

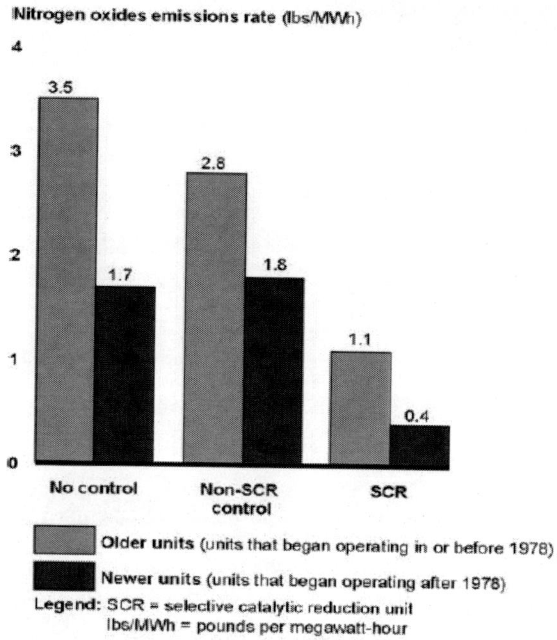

Source: GAO analysios of Ventyx data.

Note: "Non-SCR" refers to any other emissions control other than an SCR (see category 2 in table 2). Chart reflects generating units that use coal, natural gas, and oil. Older generating units are mostly coal fired, whereas newer generating units rely more heavily on natural gas.

Figure 12. Emissions Rates of Nitrogen Oxides for Older and Newer Fossil Fuel Generating Units in 2010, by Type of Control for Nitrogen Oxides.

As shown in figure 12, older units collectively emitted nitrogen oxides at higher rates than newer units. For example, older units without any nitrogen oxides controls emitted an average of 3.5 pounds of nitrogen oxides per megawatt-hour of electricity generated in 2010, approximately twice the rate as newer units.

The increased use of emissions controls over time may be attributable, in part, to the Clean Air Act requirements such as NSR and the Acid Rain Program. The Acid Rain Program, created by the Clean Air Act Amendments of 1990, established a cap on the amount of sulfur dioxide that may be emitted by electricity generating units nationwide. According to EPA officials, both NSR and the Acid Rain Program have contributed to more widespread use of emissions controls. In addition, EPA has recently finalized regulations that may prompt generating unit owners to (1) retrofit older units with emissions controls or (2) retire units that are not cost-effective to retrofit.[21] Conversely, some uncontrolled units may be limited in their ability to install certain types of emissions controls if these controls are not required.[22]

Older Units Are Generally Less Efficient Than Newer Units

The average efficiency with which electricity generating units produce electricity has improved over time, which has likely contributed to lower collective emissions among newer units. Generating efficiency is often measured by the total heat (fuel) required to generate each unit of electricity, known as a "heat rate."[23] More efficient units require less fuel to produce the same amount of electricity, generally resulting in lower emissions and fuel costs. A number of factors can affect a unit's efficiency, including its design, its operation, the choice of fuel, the use of emission controls, and the environmental conditions in which the unit operates. For example, some types of coal are more efficient than others because they have lower ash and water content. Emissions controls can also reduce a unit's efficiency because some electricity must be used to operate the controls.

As shown in figure 13, the operating efficiency of generating units built before 1950 is much lower than newer generating units. On average, in 2010, units that began operating in 2000 or later were approximately 46 percent more efficient than units that began operating before 1950. In addition to a greater reliance on coal, many older units only operate periodically, during times of peak electricity demand, which decreases their efficiency relative to units that operate on a more continuous basis. 24 While some energy efficiency technologies are available for application to existing units, the biggest efficiency gains result from the substitution of old units with new, more efficient units. Some natural gas units constructed in the last two decades, for example, are capable of achieving heat rates below 7,000 million British thermal units (Btu) per kilowatt-hour, which is substantially lower than the averages highlighted in figure 13.

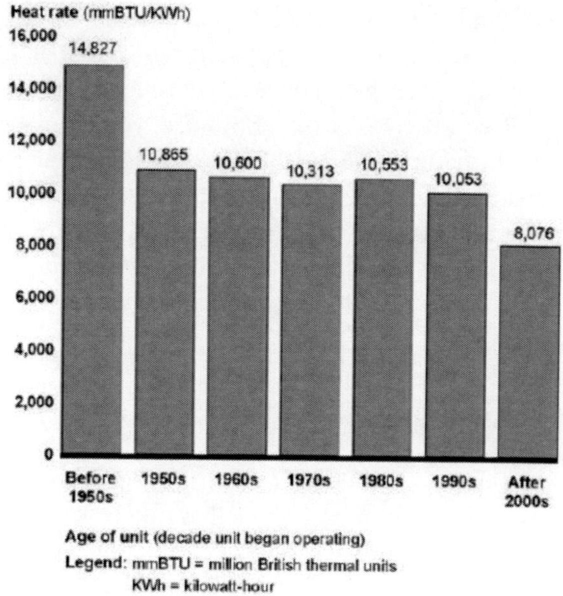

Source: GAO analysios of Ventyx data.

Note: A British thermal unit (Btu) is a standard unit of measurement used to denote the amount of heat energy in fuels. A lower heat rate indicates a higher level of efficiency. Chart reflects generating units that use coal, natural gas, and oil. Older generating units are mostly coal fired, whereas newer generating units rely more heavily on natural gas.

Figure 13. Average Efficiency (Heat Rate) of Fossil Fuel Units in 2010, by Decade Unit Began Operating.

Nonetheless, older units remain an important part of the U.S. electricity generating sector. In 2010, older units were responsible for 45 percent of total electricity production from fossil fuel units. In certain regions, older units played a more significant role in providing electricity. For example, in the Mid-Atlantic and Great Lakes regions, older units generated 64 percent and 70 percent, respectively, of electricity coming from fossil fuel units.[25] Some older units also provide services that help ensure the reliable flow of electricity to certain regions; for example, some older units may be used to help restart the electricity system in the event of a blackout.

AGENCY COMMENTS

We provided a draft of this report to EPA and the Department of Energy for their review and comment. EPA and the Department of Energy provided technical comments, which we have incorporated as appropriate.

Sincerely yours,
David C. Trimble
Director, Natural Resources and Environment
Frank Rusco
Director, Natural Resources and Environment

ENCLOSURE I. ELECTRICITY-GENERATION AND AGGREGATE EMISSIONS FROM FOSSIL FUEL ELECTRICITY GENERATING UNITS IN 2010, BY STATE

Table 3 presents, by state, data on fossil fuel units' electricity generation and aggregate emissions of sulfur dioxide, nitrogen oxides, and carbon dioxide in 2010.

Table 3. Electricity and Emissions from Fossil Fuel Electricity Generating Units in 2010, by State

State	Number of units	Electricity generation (megawatt-hours)	Carbon dioxide emissions (tons)	Nitrogen oxide emissions (tons)	Sulfur dioxide emissions (tons)
Alabama	88	100,961,300	84,828,993	63,336	204,195
Arizona	63	74,817,174	60,732,544	60,524	36,445
Arkansas	30	40,295,345	36,271,855	37,783	67,084
California	179	80,854,400	36,929,068	4,467	226
Colorado	58	45,728,202	44,826,939	54,088	45,391
Connecticut	39	12,497,043	8,526,856	2,783	1,955
Delaware	15	5,489,585	4,318,918	4,330	14,496
District of Columbia	2	207,916	220,765	373	874
Florida	246	188,566,114	135,313,627	79,759	144,829

Table 3. (Continued)

State	Number of units	Electricity generation (megawatt-hours)	Carbon dioxide emissions (tons)	Nitrogen oxide emissions (tons)	Sulfur dioxide emissions (tons)
Georgia	141	95,973,778	87,569,772	60,580	218,906
Idaho	7	1,566,178	693,069	113	3
Illinois	196	97,862,368	107,058,198	77,063	220,077
Indiana	126	120,785,560	124,321,920	121,844	414,638
Iowa	48	41,249,495	45,296,689	44,796	104,389
Kansas	48	35,758,838	39,757,268	48,938	45,251
Kentucky	96	96,980,846	101,053,094	93,040	270,045
Louisiana	70	70,221,777	52,344,198	47,153	102,263
Maine	11	7,902,550	3,953,551	719	821
Maryland	46	26,381,736	27,999,389	19,434	29,946
Massachusetts	48	31,174,655	19,787,576	8,181	36,892
Michigan	96	76,833,143	74,272,141	79,566	242,409
Minnesota	45	31,496,681	32,900,506	29,911	41,076
Mississippi	57	40,802,413	30,624,811	29,773	54,696
Missouri	102	77,566,300	83,182,289	58,288	235,368
Montana	12	18,752,877	21,356,366	21,728	19,896
Nebraska	29	23,688,902	26,402,103	37,402	64,184
Nevada	37	28,136,479	17,222,084	10,812	7,889
New Hampshire	7	8,069,450	5,899,639	4,612	36,833
New Jersey	107	30,081,377	19,995,676	9,937	15,270
New Mexico	35	34,294,657	32,657,348	60,256	16,570
New York	165	63,663,802	41,687,271	28,311	47,488
North Carolina	104	78,576,396	76,242,017	53,390	120,254
North Dakota	11	28,269,255	33,609,781	54,744	124,096
Ohio	135	126,505,558	125,054,961	104,871	572,110
Oklahoma	66	60,394,162	49,876,813	71,434	85,135
Oregon	8	19,280,743	10,875,349	9,729	15,696
Pennsylvania	123	143,267,035	125,643,066	134,169	411,216
Rhode Island	8	7,094,295	3,504,392	578	18
South Carolina	71	48,054,008	44,815,737	28,123	94,641
South Dakota	8	3,255,521	3,765,854	12,433	12,589
Tennessee	85	45,166,515	46,449,221	31,239	118,723
Texas	284	318,045,099	255,708,493	145,798	461,754
Utah	26	39,211,138	37,005,962	61,415	21,598
Virginia	97	42,773,143	35,584,177	38,300	93,385
Washington	13	17,402,607	14,560,232	11,980	2,651
Wisconsin	84	46,382,410	49,753,419	33,082	109,158
West Virginia	50	79,543,178	79,135,925	53,757	109,028
Wyoming	21	42,829,664	49,382,465	60,902	64,849
Total	3,443	2,754,711,668	2,448,972,387	2,105,844	5,157,306

Source: GAO presentation of Ventyx data.

Note: As with most of the other data presented in this report, we limited our analysis to those units that (1) list a fossil fuel as a primary fuel, (2) have a net summer capacity greater than 25 megawatts, and (3) produced power in 2010. No reported electricity production or emissions were reported from fossil fuel units in Alaska, Hawaii, or Vermont.

ENCLOSURE II. ELECTRICITY GENERATION AND EMISSIONS RATES FROM FOSSIL FUEL ELECTRICITY GENERATING UNITS IN 2010, BY STATE

Table 4 presents, by state, data on fossil fuel units' electricity generation and emissions rates in 2010.

Table 4. Electricity Generation and Emissions Rates from Fossil Fuel Electricity Generating Units in 2010, by State

State	Number of units	Electricity generation (MWh)	Carbon dioxide emissions rate (tons/MWh)	Nitrogen oxides emissions rate (lbs/MWh)	Sulfur dioxide emissions rate (lbs/MWh)
Alabama	88	100,961,300	0.84	1.25	4.05
Arizona	63	74,817,174	0.81	1.62	0.97
Arkansas	30	40,295,345	0.90	1.88	3.33
California	179	80,854,400	0.46	0.11	0.01
Colorado	58	45,728,202	0.98	2.37	1.99
Connecticut	39	12,497,043	0.68	0.45	0.31
Delaware	15	5,489,585	0.79	1.58	5.28
District of Columbia	2	207,916	1.06	3.59	8.41
Florida	246	188,566,114	0.72	0.85	1.54
Georgia	141	95,973,778	0.91	1.26	4.56
Idaho	7	1,566,178	0.44	0.14	0.00
Illinois	196	97,862,368	1.09	1.57	4.50
Indiana	126	120,785,560	1.03	2.02	6.87
Iowa	48	41,249,495	1.10	2.17	5.06
Kansas	48	35,758,838	1.11	2.74	2.53
Kentucky	96	96,980,846	1.04	1.92	5.57
Louisiana	70	70,221,777	0.75	1.34	2.91
Maine	11	7,902,550	0.50	0.18	0.21
Maryland	46	26,381,736	1.06	1.47	2.27
Massachusetts	48	31,174,655	0.63	0.52	2.37
Michigan	96	76,833,143	0.97	2.07	6.31
Minnesota	45	31,496,681	1.04	1.90	2.61
Mississippi	57	40,802,413	0.75	1.46	2.68
Missouri	102	77,566,300	1.07	1.50	6.07

Table 4. (Continued)

State	Number of units	Electricity generation (MWh)	Carbon dioxide emissions rate (tons/MWh)	Nitrogen oxides emissions rate (lbs/MWh)	Sulfur dioxide emissions rate (lbs/MWh)
Montana	12	18,752,877	1.14	2.32	2.12
Nebraska	29	23,688,902	1.11	3.16	5.42
Nevada	37	28,136,479	0.61	0.77	0.56
New Hampshire	7	8,069,450	0.73	1.14	9.13
New Jersey	107	30,081,377	0.66	0.66	1.02
New Mexico	35	34,294,657	0.95	3.51	0.97
New York	165	63,663,802	0.65	0.89	1.49
North Carolina	104	78,576,396	0.97	1.36	3.06
North Dakota	11	28,269,255	1.19	3.87	8.78
Ohio	135	126,505,558	0.99	1.66	9.04
Oklahoma	66	60,394,162	0.83	2.37	2.82
Oregon	8	19,280,743	0.56	1.01	1.63
Pennsylvania	123	143,267,035	0.88	1.87	5.74
Rhode Island	8	7,094,295	0.49	0.16	0.01
South Carolina	71	48,054,008	0.93	1.17	3.94
South Dakota	8	3,255,521	1.16	7.64	7.73
Tennessee	85	45,166,515	1.03	1.38	5.26
Texas	284	318,045,099	0.80	0.92	2.90
Utah	26	39,211,138	0.94	3.13	1.10
Virginia	97	42,773,143	0.83	1.79	4.37
Washington	13	17,402,607	0.84	1.38	0.30
West Virginia	50	79,543,178	0.99	1.35	2.74
Wisconsin	84	46,382,410	1.07	1.43	4.71
Wyoming	21	42,829,664	1.15	2.84	3.03
Total	3,443	2,754,711,668	-	-	-

Source: GAO presentation of Ventyx data.

Legend

Lb = pound

MWh = megawatt-hour

Note: As with most of the other data presented in this report, we limited our analysis to those units that (1) list a fossil fuel as a primary fuel, (2) have a net summer capacity greater than 25 megawatts, and (3) produced power in 2010. No reported electricity production or emissions were reported from fossil fuel units in Alaska, Hawaii, or Vermont.

ENCLOSURE III. ENERGY AND EMISSIONS FROM U.S. FOSSIL FUEL ELECTRICITY GENERATING UNITS, BY LOCATION IN 2010

Figures 14,15, 16, and 17 show the location of electricity generating units and the amount of electricity, sulfur dioxide, nitrogen oxides, and carbon dioxide they produced in 2010. Units in the Great Lakes, South Central, and Southeast regions produced most of the emissions.[26]

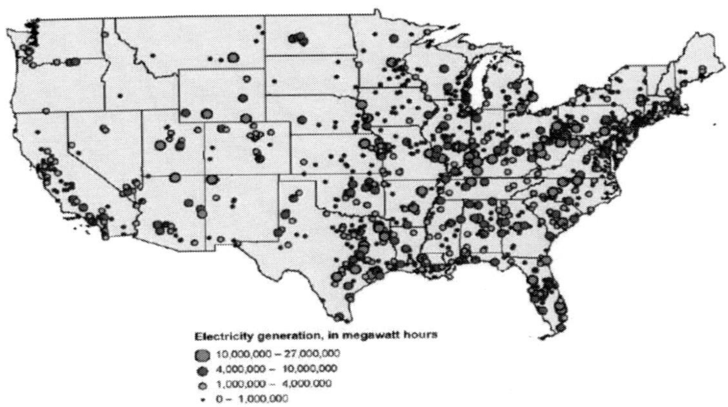

Source: GAO analysios of Ventyx data.

Figure 14. Geographic Distribution of Electricity Generation from Fossil Fuel Electricity Generating Units, 2010.

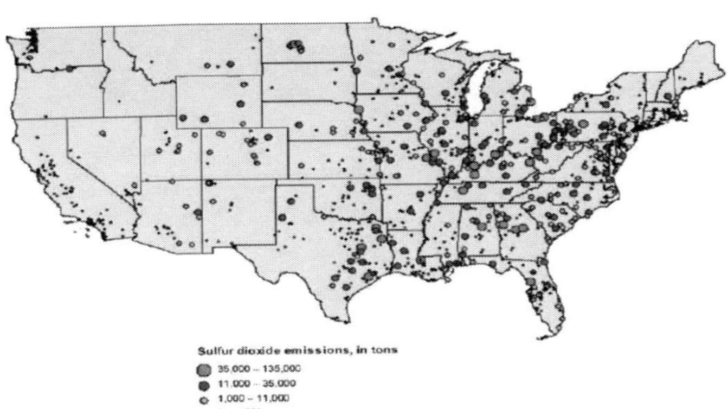

Source: GAO analysios of Ventyx data.

Figure 15. Geographic Distribution of Sulfur Dioxide Emissions from Fossil Fuel Electricity Generating Units, 2010.

Specifically, in 2010, units in these regions accounted for 69 percent of the sulfur dioxide, 59 percent of nitrogen oxides, and 63 percent of the carbon dioxide emitted from older units nationwide in 2010 and generated 62 percent of the elecricity from fossil fuel units.

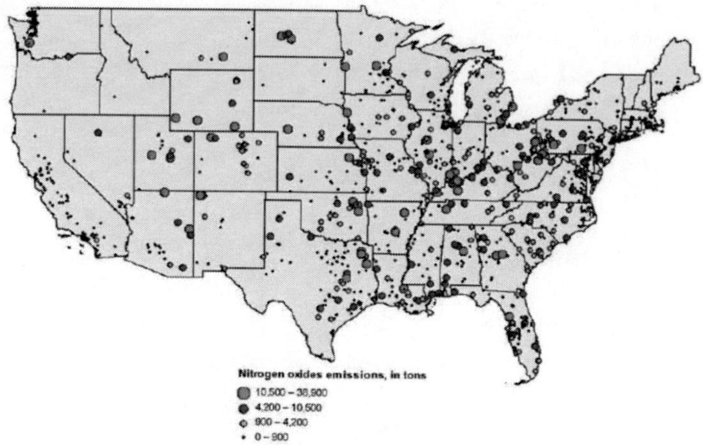

Source: GAO analysios of Ventyx data.

Figure 16. Geographic Distribution of Nitrogen Oxides Emissions from Fossil Fuel Electricity Generating Units, 2010.

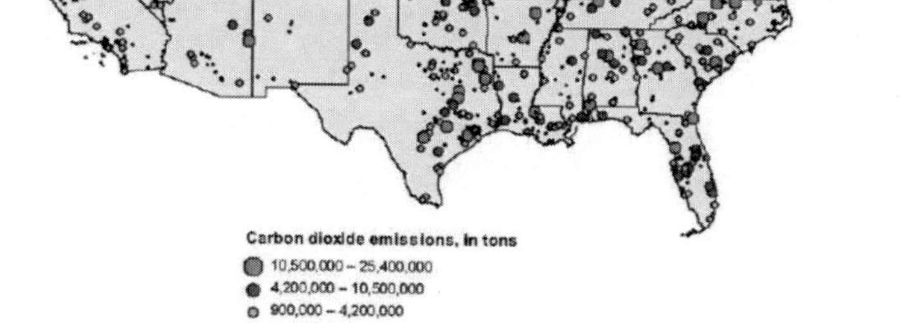

Source: GAO analysios of Ventyx data.

Figure 17. Geographic Distribution of Carbon Dioxide Emissions from Fossil Fuel Electricity Generating Units, 2010.

ENCLOSURE IV. ENERGY AND EMISSIONS FROM U.S. FOSSIL FUEL ELECTRICITY GENERATING UNITS IN 2010, BY DECADE UNIT BEGAN OPERATING

Table 5 presents data on fossil fuel units' electricity generation, aggregate emissions, capacity, and emissions per unit of electricity generated, based on the decade the unit began operating.

Table 5. Electricity Generation and Emissions in 2010, by Decade Unit Began Operating

			Decade unit began operating			
	1950s	**1960s**	**1970s**	**1980s**	**1990s**	**2000s**
Number of units	415	382	683	232	404	1,249
Electricity generation (MWh, in thousands)	195,561	353,497	770,451	513,904	163,507	728,165
Net summer capacity (MW)	52,130	93,570	177,014	83,160	52,967	241,798
Carbon dioxide emissions (tons, in thousands)	212,070	370,819	799,849	551,016	118,900	369,887
Carbon dioxide emissions rate (lbs/MWh)	2,169	2,098	2,076	2,144	1,454	1,016
Nitrogen oxides emissions (lbs)	273,290	446,312	697,687	524,769	85,282	65,186
Nitrogen oxides emissions rate (lbs/MWh)	2.79	2.53	1.81	2.04	1.04	0.18
Sulfur dioxide emissions (lbs)	1,135,001	1,372,766	1,506,432	987,113	74,822	31,368
Sulfur dioxide emissions (lbs/MWh)	11.61	7.77	3.91	3.84	0.92	0.09

Source: GAO analysis of Ventyx data.

Legend

lb = pound

MWh = megawatt-hour.

End Notes

[1] Fossil fuels are responsible for nearly all emissions of carbon dioxide, sulfur dioxide, and nitrogen oxides from power plants.

[2] An electricity generating unit consists of any combination of an electricity generator, reactor, boiler, combustion turbine, or other equipment operated together to produce electrical power. A power plant is a facility with one or more generating units, together with other equipment used to produce electric power.

[3] Oil is used to a very limited extent, providing less than 1 percent of electricity in 2010.

[4] EPA has set national ambient air quality standards for six pollutants, termed "criteria" pollutants: carbon monoxide, lead, nitrogen oxides, ozone, particulate matter, and sulfur oxides.

[5] This report focuses solely on fossil fuel electricity generating units, although NSR also applies to certain other major stationary sources of air pollution, such as other industrial facilities.

[6] We used 1978 as the cutoff date for our analysis—instead of August 7, 1977—because 1978 was the first full year after NSR's establishment, and our analysis reports only annual data.

[7] GAO, *Air Pollution: Emissions from Older Electricity Generating Units*, GAO-02-709 (Washington, D.C.: June 12, 2002). Note: This report defines "older units" as those fossil fuel generating units that began operation before 1972.

[8] Regulated pollutants are sulfur oxides; nitrogen oxides; other criteria pollutants; and, since January 2, 2011, greenhouse gases.

[9] EPA, *National Emissions Inventory Air Pollutant Emissions Trends Data*, "Current Emissions Trends Summaries," accessed October 2011, http://www.epa.gov/ttn/chief/ trends/index. html#tables.

[10] EPA, Inventory of U.S. Greenhouse Gas Emissions and Sinks: 1990-2009, EPA-430-R-11-005 (Washington, D.C.: Apr. 15, 2011).

[11] GAO-02-709 and GAO, Air Quality: Information on Tall Smokestacks and Their Contribution to Interstate Transport of Air Pollution, GAO-11-473 (Washington, D.C.: May 11, 2011).

[12] Sulfur dioxide and nitrogen oxides can transform into fine particles in the atmosphere. Fine particles are a subset of particulate matter, a regulated pollutant. Ozone, also a regulated pollutant, forms when nitrogen oxides react with volatile organic compounds (chemicals that are emitted as gases from natural sources, such as trees, as well as from anthropogenic sources, such as motor vehicles) in the presence of heat and sunlight.

[13] GAO-11-473.

[14] EPA, Office of Air and Radiation, Available and Emerging Technologies for Reducing Greenhouse Gas Emissions from Coal-Fired Electric Generating Units (Research Triangle Park, NC: October 2010).

[15] EPA, *Integrated Planning Model,* "TR Base Case v.4.10," accessed April 2, 2012, http://www.epa.gov/airmarkets/progsregs/epa-ipm/BaseCasev410.html#documentation.

[16] Of the 3,443 fossil fuel electricity generating units we assessed, 1,485 were older units, and 1,958 were newer units.

[17] In 2010, oil-fired units (both older and newer) accounted for less than one percent of electricity from fossil fuel units.

[18] The 240 percent increase refers to 1987 through 2009. See Energy Information Administration, *Annual Energy Review 2010*, DOE/EIA-0384(2010) (Washington, D.C.: October 2011).

[19] GAO, Natural Gas: Factors Affecting Prices and Potential Impacts on Consumers, GAO-06-420T (Washington, D.C.: Feb. 13, 2006).

[20] We limited this analysis to coal-fired units because these units are responsible for nearly all (99 percent) of sulfur dioxide emissions from fossil fuel units.

[21] These regulations include the Cross-State Air Pollution Rule—which limits sulfur dioxide and nitrogen oxides emissions from a number of states that contribute significantly to nonattainment of or interference with maintenance of certain national ambient air quality standards in downwind states—and National Emissions Standards for Hazardous Air Pollutants from Coal- and Oil-Fired Electric Utility Steam Generating Units, also known as the Mercury and Air Toxics Standards, which establish emissions limitations on mercury and other pollutants. Both regulations are being challenged in court.

[22] In many states, public utility commissions are responsible for approving the rates (or prices) electric utilities charge their customers and ensuring they are reasonable. As part of approving rates, these commissions approve such utility investments as emissions controls and, as a result, may consider whether specific technologies are reasonable. These commissions generally require some demonstration that the investment is prudent.

[23] Heat rate is a measurement used in the energy industry to calculate how efficiently an electricity generator uses heat energy. It is expressed as the number of British thermal units (Btu) of heat required to produce a kilowatt hour of energy. A lower heat rate signifies greater efficiency.

[24] Generating units that undergo frequent startups use comparatively large amounts of fuel and may be less efficient than units that operate continuously.

[25] These regions correspond to Environmental Protection Agency (EPA) Region 3 (Delaware, District of Columbia, Maryland, Pennsylvania, Virginia, and West Virginia) and Region 5 (Illinois, Indiana, Michigan, Minnesota, Ohio, and Wisconsin).

[26] These regions correspond to Environmental Protection Agency (EPA) Region 5 (Illinois, Indiana, Michigan, Minnesota, Ohio, and Wisconsin); Region 6 (Arkansas, Louisiana, New Mexico, Oklahoma, and Texas); and Region 4 (Alabama, Florida, Georgia, Kentucky, Mississippi, North Carolina, South Carolina, and Tennessee).

In: Emissions Control in Electricity Generation ISBN: 978-1-62257-830-6
Editors: J. L. Walker, A. Hernandez

Chapter 4

AIR POLLUTION: EPA NEEDS BETTER INFORMATION ON NEW SOURCE REVIEW PERMITS*

United States Government Accountability Office

WHY GAO DID THIS STUDY

Electricity generating units that burn fossil fuels supply most of the nation's electricity and are major sources of air pollution. Under the Clean Air Act, such units are subject to NSR, a permitting process that applies to (1) units built after August 7, 1977, and (2) existing units that undertake a major modification. Owners of such units must obtain from the appropriate permitting agency a preconstruction permit that sets emission limits and requires the use of certain pollution control technologies. EPA oversees states' implementation of NSR, including reviewing and commenting on draft permits issued by state and local permitting agencies. GAO was asked to examine (1) what information EPA maintains on NSR permits issued to fossil fuel electricity generating units; (2) challenges, if any, that EPA, state, and local agencies face in ensuring compliance with requirements to obtain NSR permits; and (3) what available data show about compliance with requirements

* This is an edited, reformatted and augmented version of the Highlights of GAO-12-590, a report to the Chairman, Subcommittee on Oversight, Committee on Environment and Public Works, U.S. Senate, dated June 2012.

to obtain NSR permits. GAO reviewed relevant documentation and interviewed EPA, state, and local officials, as well as representatives from industry, research, and environmental groups.

What GAO Recommends

GAO recommends that EPA, among other actions, consider ways to develop a centralized source of data on NSR permits issued to electricity generating units. EPA expressed its commitment to filling gaps in its data systems, but disagreed with the actions GAO recommended. GAO believes that its recommendations would enhance oversight of NSR permitting and enforcement.

What GAO Found

The Environmental Protection Agency (EPA) does not maintain complete information on New Source Review (NSR) permits issued to fossil fuel electricity generating units. State and local permitting agencies track the NSR permits they issue, but EPA does not maintain complete or centralized information on permits, despite a 2006 recommendation by the National Research Council that it do so. EPA maintains several databases that compile data on draft and issued NSR permits, but these sources are incomplete and thus cannot be used to identify all of the NSR permits that have been issued nationwide. In addition, EPA has the opportunity to review and comment on every draft NSR permit issued by state and local permitting agencies, but it does not compile data on whether the permitting agencies address EPA's comments in final permits. The absence of more complete information on NSR permitting makes it difficult to know which units have obtained NSR permits or to assess how state and local permitting agencies vary from EPA in their interpretations of NSR requirements.

Officials from EPA, state, and local agencies face challenges in ensuring that owners of fossil fuel electricity generating units comply with requirements to obtain NSR permits. Many of these challenges stem from two overarching issues. First, in some cases it is difficult to determine whether an NSR permit is required. NSR applicability depends on, among other factors, whether a change to a unit qualifies as routine maintenance, repair, and replacement; and whether the change results in a significant net increase in emissions. The rules governing NSR are complex, however, and applicability is determined on a case-by-case basis. Second, it is often difficult to identify noncompliance—instances where unit owners made a major modification without first obtaining an NSR permit—partly because owners of generating units determine whether a permit is needed, and in many cases their determinations are not reviewed by permitting agencies or EPA. State permitting agencies generally issue NSR permits, but EPA typically leads enforcement efforts, since identifying instances of noncompliance involves extensive investigations that go beyond the routine inspections conducted by state and local permitting agencies. EPA identifies NSR noncompliance through a lengthy, resource-intensive process that involves reviewing large amounts of information on units' past emissions and construction activities.

Available data on compliance, although incomplete, suggest that a substantial number of generating units did not comply with requirements to obtain NSR permits. Complete NSR compliance data do not exist for two main reasons: (1) EPA has not yet investigated all generating units for compliance, and (2) NSR compliance is determined at a point in time, and in some cases federal courts have disagreed with EPA about the need for an NSR permit. Nonetheless, EPA has investigated most coal-fired generating units at least once, and has alleged noncompliance at more than half of the units it investigated. Specifically, of the 831 units EPA investigated, 467 units were ultimately issued notices of violation, had complaints filed in court, or were included in settlement agreements. In total, EPA reached 22 settlements covering 263 units, which will require affected unit owners to, among other things, install around $12.8 billion in emissions controls. These settlements will reduce emissions of sulfur dioxide by an estimated 1.8 million tons annually, and nitrogen oxides by an estimated 596,000 tons annually.

ABBREVIATIONS

DOJ	Department of Justice
EIA	Energy Information Administration
EPA	Environmental Protection Agency
NSR	New Source Review
OECA	Office of Enforcement and Compliance Assurance
SCR	selective catalytic reduction

June 22, 2012

The Honorable Sheldon Whitehouse
Chairman
Subcommittee on Oversight
Committee on Environment and Public Works
United States Senate

Dear Mr. Chairman:

The efficient and reliable operation of the electricity industry is critical to the health of the U.S. economy. Residential consumers rely on electricity to power their households, and electricity is a key input for businesses that produce trillions of dollars in products and services. The United States depends on a variety of fuels to generate electricity, including fossil fuels (coal, natural gas, and oil), renewable sources (such as wind, solar, and biomass), and nuclear power. Of these, fossil fuels have historically been the primary source of U.S. electricity because of their abundance, reliability, and relatively low cost. Yet in addition to providing about 70 percent of U.S. electricity, electricity generating units at fossil fuel power plants produce substantial amounts of harmful air emissions.[1] In particular, fossil fuel-fired electricity generating units are among the largest emitters of sulfur dioxide and nitrogen oxides, which have been linked to respiratory illnesses and acid rain, as well as of carbon dioxide, the primary greenhouse gas contributing to climate change.

Under the Clean Air Act, the Environmental Protection Agency (EPA) establishes national ambient air quality standards for six pollutants which states, primarily, are responsible for attaining.[2] States attain these standards, in part, by regulating emissions of these pollutants from certain stationary sources, such as electricity generating units. Numerous Clean Air Act requirements apply to electricity generating units, including New Source Review (NSR), a permitting process established in 1977.[3] Under NSR, owners of generating units must obtain a preconstruction permit that establishes emission limits and requires the use of certain emissions control technologies.[4] NSR applies to (1) generating units built after August 7, 1977, and (2) existing generating units—regardless of the date built—that seek to undertake a "major modification," a physical or operational change that would result in a significant net increase in emissions of a regulated pollutant.[5] Units built before August 7, 1977, are not required to undergo NSR unless they undertake a major modification. However, EPA's regulatory definition of major modification excludes certain activities; generating units can undertake these activities without obtaining NSR permits or installing any additional controls. For example, activities that qualify as "routine maintenance, repair, and replacement" are not considered major modifications and therefore do not trigger NSR.

Congress allowed units that existed as of August 7, 1977 to defer installation of emissions controls until they made a major modification in the expectation that, over time, all units would either install such controls or shut down, thereby lowering overall emissions. As we reported in April 2012, however, many older units—those that began operating in or before 1978—continue to produce electricity.[6] According to our analysis of EPA data, 1,485 older units (43 percent of fossil fuel units) were still in operation in 2010.[7] A substantial number of these older units did not have emissions controls; for example, 1,201 units (74 percent of older units) did not have controls for sulfur dioxide. In addition, 564 units (38 percent of older units) did not have any controls for nitrogen oxides, and 1,277 units (86 percent of older units) had not installed selective catalytic reduction (SCR) equipment, the type of control most effective at reducing nitrogen oxides. The reduction in emissions from the use of emissions controls can be substantial: SCR equipment, for example, can reduce nitrogen oxides emissions by 70 to 95 percent. Emissions controls can also be expensive; for example, installing SCR equipment in a typical coal-fired generating unit could cost from $108 million to $129 million.

Throughout its history, NSR has been characterized by complexity and controversy, involving disputes between EPA and industry about, among other issues, whether certain changes to generating units qualified for exclusion as routine maintenance, repair, and replacement. In December 2002, EPA finalized revisions to NSR regulations, including exemptions for certain pollution control projects from NSR. These revisions were intended to maximize operating flexibility, improve environmental quality, and promote administrative efficiency, among other aims. In addition, in October 2003, EPA finalized a rule that categorically excluded certain activities from NSR by defining them as "routine maintenance, repair, and replacement." This rule was intended to provide more certainty to generating units and permitting authorities. These NSR reforms, as the two rulemakings were known, provoked considerable controversy. Some states and industry groups agreed with EPA's position, while a number of public health and environmental groups, as well as a group of states primarily from the Mid-Atlantic and

Northeast, filed lawsuits challenging the legality of the two rules. Since their issuance, two of the five provisions in the 2002 rule, as well as the 2003 rule, were struck down in court.[8]

As with many environmental laws, responsibility for implementing NSR, including issuing NSR permits, generally rests with state and local agencies, with oversight by EPA's 10 regional offices and EPA headquarters. Owners or operators of generating units wishing to make major modifications must prepare and submit a permit application to the appropriate state or local permitting agency before construction. The state or local permitting agency determines if the application is complete; develops a draft permit, if one is necessary; notifies EPA and the public of the application; and solicits comments on the draft permit. The permitting agency then issues a final permit, if merited, with responses to comments it received. The permitting agency is to provide EPA with a copy of every permit application, draft permit, and final permit.

EPA administers its NSR oversight and enforcement responsibilities through its headquarters Office of Enforcement and Compliance Assurance, as well as its 10 regional offices. This office monitors compliance, identifies national enforcement concerns and priorities, and provides overall direction on enforcement policies; the office also occasionally takes enforcement action in conjunction with the Department of Justice (DOJ), the agency responsible for handling enforcement cases in federal court. In turn, EPA's regional offices carry out much of EPA's NSR enforcement responsibilities, oversee states' enforcement programs, and implement NSR in certain areas, such as Indian country.[9] Since 1999, EPA has made enforcing NSR among coal-fired electricity generating units a national enforcement priority and has reached several settlements with owners of such units, which have resulted in the installation of emissions controls, unit retirements, agreements to fund environmentally beneficial projects, and tens of millions of dollars in civil penalties.

This report responds in part to your request for information on the implementation and enforcement of NSR for fossil fuel electricity generating units. Specifically, our objectives were to examine (1) what information EPA maintains on NSR permits issued to fossil fuel electricity generating units; (2) challenges, if any, that EPA, state, and local agencies face in ensuring compliance with requirements to obtain NSR permits; and (3) what available data show about compliance with requirements to obtain NSR permits.

To respond to our objectives, we gathered information from EPA and from selected state and local officials involved in implementing NSR. We selected a nonprobability sample of nine states on the basis of (1) the number of older electricity generating units in the state; (2) the quantity of electricity generated by such units in those states; (3) the volume of sulfur dioxide, nitrogen oxides, and carbon dioxide emitted by units in those states; and (4) the region in which the generating unit was located. The nine states we selected were: Alabama, Georgia, Indiana, Kentucky, Missouri, New York, North Carolina, Ohio, and Pennsylvania. We gathered information from EPA and these selected states on the status of their NSR permitting programs and efforts to collect and maintain relevant data. In addition, we spoke with officials from these states, as well as officials at the four EPA regional offices that oversee these states and EPA's Office of Air and Radiation. In three of the states, some localities are responsible for NSR permitting; we spoke with officials at two of those localities. To assess compliance with NSR, we reviewed relevant documents from EPA, other government agencies, academic and research institutions, environmental organizations, and industry groups. We also interviewed knowledgeable enforcement and compliance officials from EPA headquarters and the four regional offices, as well as officials in selected states and localities, to collect information on EPA's efforts to enforce NSR and track compliance among generating units. In addition, we spoke with stakeholders from academic and research institutions, environmental organizations, and industry groups on these topics. Appendix I provides additional information about our scope and methodology.

We conducted this performance audit from April 2011 to June 2012, in accordance with generally accepted government auditing standards. Those standards require that we plan and perform the audit to obtain sufficient, appropriate evidence to provide a reasonable basis for our findings and conclusions based on our audit objectives. We believe that the evidence obtained provides a reasonable basis for our findings and conclusions based on our audit objectives.

BACKGROUND

Subject to EPA's oversight, state and local permitting agencies generally administer NSR and operate under one of two arrangements. Under the first arrangement, state and local agencies receive "delegated authority" from EPA under which they implement EPA's NSR regulations. Under the second arrangement, states and localities are also responsible for administering NSR, but instead of implementing EPA's NSR regulations, state and local agencies develop plans, known as state implementation plans, that regulate the construction and modification of stationary sources. These plans provide assurances that the states and localities will have adequate personnel, funding, and authority under state law to carry out the plan, among other provisions. State implementation plans also must include NSR regulations that are at least as stringent as EPA's NSR regulations, although states and local agencies are authorized to include more stringent or additional requirements.[10] States and localities must submit these plans, as well as any revisions to them, to EPA for approval. Once EPA approves the plans, they become federally enforceable requirements.

Although this report focuses on NSR, the Clean Air Act and its implementing regulations subject electricity generating units to additional emissions control requirements. For example, the Acid Rain Program, created by the Clean Air Act Amendments of 1990, established a cap on the amount of sulfur dioxide that may be emitted by electricity generating units nationwide and authorizes those generating units to trade emissions allowances for sulfur dioxide. These facilities must also continuously monitor their emissions and report them to EPA. Furthermore, EPA has recently finalized or proposed several other regulations that will affect many fossil fuel generating units. These regulations include the (1) Mandatory Reporting of Greenhouse Gas rule finalized in 2009, which established reporting requirements for greenhouse gas emissions above certain thresholds; (2) Cross-State Air Pollution Rule, finalized in 2011, which limits sulfur dioxide and nitrogen oxides emissions from a number of states that contribute significantly to nonattainment or interference with maintenance of certain national ambient air quality standards in downwind states; (3) National Emissions Standards for Hazardous Air Pollutants from Coal- and Oil-Fired Electric Utility Steam Generating Units, also known as the Mercury and Air Toxics Standards, which establish emissions limitations on mercury and other pollutants and was finalized on February 15, 2012; and (4) Standards of Performance for Greenhouse Gas Emissions for New Stationary Sources for Electric Utility Generating Units, proposed in April 2012, which establishes new source performance standards for emissions of carbon dioxide for certain new fossil fuel electricity generating units.[11]

EPA DOES NOT MAINTAIN COMPLETE INFORMATION ON ISSUED NSR PERMITS OR TRACK THE IMPACT OF ITS COMMENTS

EPA does not maintain complete information on NSR permits issued to fossil fuel electricity generating units. State and local permitting agencies track the NSR permits they issue, but EPA does not maintain data on these permits in a complete and centralized source of information, which limits the agency's ability to assess the impact of NSR. In addition, EPA has the opportunity to review and comment on every draft NSR permit issued by state and local permitting agencies, but the agency does not compile data on which permitting authorities address EPA's comments. The absence of this information makes it difficult for EPA to measure the impact of its comments and may impede its ability to assess how state and local permitting agencies may differ from EPA in their interpretation of NSR requirements.

EPA Does Not Maintain Complete Data on NSR Permits

EPA does not maintain complete information on NSR permits issued for construction of new fossil fuel electricity generating units or for major modifications to existing units. State and local permitting agencies, which issue NSR permits in most parts of the country, track the NSR permits they issue. (Figure 1 describes the roles of state and local permitting agencies and EPA in issuing NSR permits.) State and local agencies vary widely in the types of data they collect on NSR permits and the systems they use to compile the data. Some states maintain detailed information on NSR permits in electronic form available on publicly accessible websites. For instance, in seven of the nine states where we conducted interviews, state officials maintain information online that can be used to identify the electricity generating units that have received NSR permits, as well as the requirements of the permits.[12] These data, however, are maintained in different formats across these states and cannot be readily compiled into a complete source of information on NSR permitting for the electricity generating sector.

In addition to the data collected by state and local permitting agencies, EPA maintains several sources of data on NSR permitting for electricity generating units, but none of these sources are complete. Of the four EPA regional offices we assessed, three offices consolidated NSR permitting data for the states and localities under their oversight.[13] For example, one EPA regional office maintains a spreadsheet to track NSR permits issued over time. The spreadsheet contains information on (1) the type of source applying for the permit, (2) the project under consideration, (3) when a final permit is issued, and (4) what emissions control technologies are required. However, the information maintained by the EPA regional offices is in different formats and is not compiled into a complete, centralized repository of information on NSR permits issued nationwide.

In 2006, the National Research Council recommended that EPA, in conjunction with states and localities, systematically collect data on the full range of NSR permits issued nationwide to stationary sources of air pollution, including electricity generating units.[14] In its report the council noted that such information could be useful to stakeholders and decision makers for assessing the broader impacts of NSR regulations, or potential changes to the regulations, on emissions, public health, and energy efficiency. Asked how EPA had responded to this recommendation, officials in EPA's Office of Air and Radiation said it had largely been satisfied by an agency database that tracks the emissions control technologies that have been required by issued NSR permits.[15] However, the database is incomplete; EPA officials said it contains information on approximately 50 percent of issued NSR permits and therefore cannot be used, as recommended by the council, to identify all NSR permits that have been issued nationwide. EPA officials suggested that the database was not complete because some state and local permitting agencies do not enter their information in a timely manner, and others do not report the information at all. In addition, EPA officials noted that the Clean Air Act requires only that states enter permit information for sources in nonattainment areas.[16]

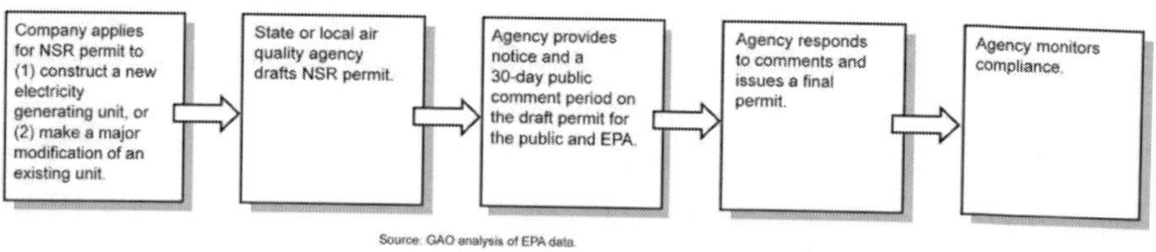

Figure 1. New Source Review Permitting Process.

EPA also has a centrally maintained database on NSR permits issued nationwide for sources of greenhouse gases. EPA officials told us this database would capture most NSR permitting activity for electricity generating units because such units are major sources of greenhouse gases, in addition to other pollutants. But this database too is incomplete: it has been used only since 2011 and does not contain information on whether the NSR permits include requirements for pollutants other than greenhouse gases. EPA officials mentioned other means of tracking data on NSR permits, but none of these sources can definitively identify the units that have received NSR permits nationwide (see app. II for more information on sources of data on NSR permits).[17]

Because EPA does not maintain comprehensive information on the NSR permits issued to fossil fuel electricity generating units across the nation, the agency does not have complete data on which units have obtained permits, limiting its ability to assess the scope and impact of NSR. Without such data, EPA cannot fully assess what controls have been required or estimate what emissions from generating units may have been averted as a result of NSR requirements.

In addition, the absence of information on NSR permits hinders EPA's ability to efficiently target noncompliance. To identify potential targets for NSR enforcement actions—units that may have made a physical or operational change that resulted in a significant net emissions increase without obtaining a permit—EPA officials we interviewed said they have relied in part on continuous emissions monitoring data required under Title IV of the Clean Air Act. For example, an unusual emissions decrease followed by a sharp increase could signal a temporary plant shutdown to install new equipment. In instances where noncompliance has been suspected, EPA officials we spoke with said they have sometimes consulted relevant state and local permitting authorities to ascertain whether a suspected unit previously received an NSR permit. In other cases, EPA officials told us they have obtained copies of NSR permits by requiring unit owners to provide them through information requests, along with documentation on previously completed work at a suspected unit—making a centralized source of information on NSR permits unnecessary. Nevertheless, without a centralized source of information on NSR permits, EPA enforcement officials cannot readily identify whether a generating unit has obtained an NSR permit, and under what terms, unless they rely on unit owners or the efforts of state and local permitting agencies to provide this information.

EPA Does Not Track Incorporation of Its Comments into Draft Permits

EPA does not track whether state and local permitting agencies incorporate the agency's comments into final NSR permits for fossil fuel electricity generating units. State and local permitting agencies have the authority to make final permitting decisions, but as part of its oversight responsibilities, EPA has the opportunity to review and comment on every draft NSR permit submitted by state and local permitting agencies.[18] EPA officials said that, when the agency elects to make comments, their reviews often focus on whether and how the permitting authorities conduct the required analyses—for example, whether an appropriate analysis was used to support a proposed emissions control as the best available control technology. The agency's comments are first prepared by regional offices and then reviewed by headquarters to ensure consistency with comments previously issued by the agency. According to EPA officials, EPA's headquarters office maintains a database to track the agency's work and ensure that it delivers comments during the 30-day public comment period. This database includes basic information on the draft permits, but EPA does not use it to track the proposed terms of the permits, the substance of the agency's own comments, whether the draft permits are finalized, or the extent to which state and local agencies incorporated EPA's comments into final permits.

Officials we interviewed in several EPA regional offices said they note the responses of state and local permitting agencies to EPA's comments, but all said that they do not regularly report this information to EPA headquarters. EPA officials also told us that regional officials notify headquarters officials, including members of a working group on the utility industry, when significant areas of concern arise on a final NSR permit for a generating unit. They noted that this information is neither reported systematically nor compiled in a centralized manner, adding that quantifying the impact of all of its comments would be subjective and time-consuming. However, without compiling any information on whether state and local permitting agencies have incorporated its comments into final NSR permits, EPA is limited in its ability to assess the impact of these comments on state and local permitting decisions across the nation. For example, EPA cannot assess the extent to which NSR emissions control requirements imposed by state and local permitting agencies may differ from those suggested by EPA in its comments.[19]

Regulators Face Challenges in Ensuring Compliance with NSR Permitting Requirements

In addition to a lack of comprehensive permitting data, EPA and state and local agencies face other challenges in ensuring that owners of fossil fuel electricity generating units comply with requirements to obtain NSR permits. Many of the challenges stem from two overarching issues: (1) determining whether an NSR permit is required and (2) identifying instances where unit owners should have obtained NSR permits but did not. As a result, EPA's enforcement efforts involve long, resource-intensive investigations.

Determining When NSR Applies Can be Difficult

A major challenge to EPA, states, and local agencies in ensuring NSR compliance is that it can be difficult for unit owners and regulators to know whether an NSR permit is needed, because NSR's rules governing applicability are complex and because NSR applicability is determined on a case-by-case basis. EPA and state officials we spoke with said that NSR as it applies to new units is fairly straightforward, because newly constructed units generally must obtain NSR permits before starting operation. In contrast, determining what constitutes a major modification of an existing unit, and, thus, what requires an NSR permit, is more complex. Under NSR regulations, owners are to apply for an NSR permit before making any physical or operational change that would result in a significant net increase of emissions. These changes, such as adding new equipment, must be evaluated in the specific context of the unit and its intended use. State officials and industry representatives we interviewed said it can be difficult to determine whether these activities trigger NSR because the two steps for determining applicability—first, whether the unit is making a physical or operational change and, second, whether this change would result in a significant net increase of emissions—are not categorically defined and have changed over time.

The first step for determining NSR applicability can be complicated because the definition of "physical or operational change" excludes activities that are considered routine maintenance, repair, and replacement. NSR regulations, first finalized in 1978, contained no description or definition of the "routine maintenance" exclusion, instead relying on a case-by-case approach that involves weighing several factors, including the nature, extent, purpose, frequency, and cost of proposed activities. Federal courts, however, have issued inconsistent decisions on whether the factors should be analyzed with respect to industry practice or a particular unit's history. In 2003, in part because of concerns about the case-by-case approach, EPA finalized a rule that categorically excluded certain activities from NSR by defining them as "routine maintenance, repair, and replacement" to provide more certainty to generating units and permitting agencies. Specifically, the rule categorically deemed certain replacement activities to be routine maintenance, repair, and replacement if certain conditions were met, such as replacement activities' costs not exceeding a specified threshold. In 2006, however, a federal appeals court struck down this rule because it was contrary to the plain language of the Clean Air Act.

As a result, a case-by-case approach is still used to determine which activities qualify for the exclusion. Several state officials and industry representatives we interviewed said that the case-by-case approach makes it difficult to know when NSR applies. A number of industry representatives also said that uncertainty around NSR applicability can deter owners from making improvements to units that would improve efficiency. One senior EPA enforcement official we interviewed, however, noted that NSR regulations are written broadly to cover many disparate industries and said it would not be possible for EPA to develop detailed regulations tailored to each industry. One state official we spoke with also said that attempts to more precisely specify what activities are considered routine maintenance might not be worthwhile, since EPA's previous efforts to do so were struck down in court.

The second step in determining NSR applicability—assessing whether a change results in a significant net increase in emissions—presents additional complications. Like the routine maintenance exclusion, regulations governing what constitutes an increase in emissions have been subject to litigation, leading to changes in the process used to measure emissions increases over time. For example, in 1992, in response to a court decision, EPA finalized a regulation changing how future emissions from generating units are to be calculated. Rather than calculating future emissions based on a unit's potential to emit, under the revised regulation, future emissions are calculated based, in part, on the maximum emissions that can be generated while operating the unit as it is intended to be operated and as it is normally operated. Some state officials and industry representatives we interviewed said that calculating emissions increases can be challenging because the regulations are complex, and EPA's interpretation has changed over time. NSR's complexity can be particularly difficult for owners of smaller generating units who may lack the legal and technical expertise to properly comply with NSR, according to an EPA official and industry representative we interviewed. EPA officials acknowledged that the process is not always simple, but they also noted that it is much easier for companies to make these calculations than for permitting agencies to verify them, since permitting agencies are less familiar with—and have less access to— information about a generating unit, its activities, and its data systems, than the companies.

According to several state officials and industry representatives we interviewed, assessing whether a change results in a significant net increase in emissions can also be complicated because EPA regulations authorize certain emissions increases to be excluded from this assessment—specifically, those emissions increases that are attributable to growth in demand.[20] Several "state officials we interviewed said that some owners have had difficulty distinguishing between emissions increases due to projected growth in demand and emissions increases resulting from the change to the unit, a process made more difficult because EPA has not offered clarification or guidance regarding this exclusion. One senior EPA enforcement official disagreed with this assessment, noting that utilities commonly employ models that help project demand as a way to guide their operations and investment decisions. According to this official, EPA's approach is based on methods already widely employed throughout the electricity sector.

EPA and state agency officials, who are responsible for verifying owners' calculations when they apply for a permit or seek guidance on NSR applicability, said that verifications are further complicated by other NSR provisions that exclude certain activities from NSR. For example, a change that significantly increases a generating unit's emissions will not trigger NSR if it does not cause a net increase in emissions. Specifically, an NSR permit is not required if the increase in emissions resulting from a change is offset by certain contemporaneous emissions reductions, a process called "netting." EPA has defined "contemporaneous" as within 5 years before construction on the change commences, although states can define the term differently. Thus, an owner could compensate for an emissions increase in a given year by subtracting emissions decreases that were made in the previous 5 years, although any other emissions increases during that 5-year period must also have been included in the calculation. Several state agency officials we spoke with said that unit owners often pursue this option so they do not have to obtain an NSR permit and install costly emissions controls. Several EPA and state officials we interviewed also said, however, that it can be difficult to verify that calculations are valid, in part because they must rely on information provided by the unit owners. Some of these officials said it can be difficult to determine what types of emissions reductions and increases may be aggregated together under the netting option.[21] One EPA regional office official said that, overall, options such as netting complicate and lengthen the permitting process because they require unit owners to submit additional documentation that the regulator must in turn review.

To aid owners and regulators in determining when NSR should apply, EPA and state officials identified several sources of available guidance, including the following:

- *Consultations with state and local agencies.* Before seeking a permit, owners of units can request assistance from state and local permitting agencies in determining whether NSR applies. Some state agency officials said that unit owners in their state regularly seek guidance, particularly on how to qualify for one of NSR's exclusions. However, other EPA and state officials we spoke with said that such requests are uncommon; many unit owners may hesitate to contact a regulatory agency because regulators may have a different interpretation of NSR that could require them to install costly emissions controls.
- *EPA's 1990 draft NSR workshop manual.* Several state agency officials we spoke with said they rely on a draft EPA manual from 1990 issued as guidance for implementing the federal NSR permitting process, although the manual was never finalized and has not been updated.
- *Regionally maintained databases.* Through one of its regional offices, EPA maintains an online database containing more than 600 EPA-issued policy and guidance documents. Several EPA and state officials we interviewed said that the database was helpful in providing current information on how to apply NSR, although one state official said that these determinations are not always consistent.
- *Court decisions.* Several EPA and state permitting officials we interviewed said they rely primarily on court rulings for guidance on interpreting NSR regulations to ensure that their determinations are up-to-date.

EPA officials said that the agency's ability to generate comprehensive, nationwide guidance is limited because of the case-by-case nature of NSR, ongoing litigation, and the variation in NSR requirements across states. For example, some states and localities have adopted NSR requirements that are more stringent than the federal regulations. Furthermore, some states' regulations differ because they have not revised their state implementation plans to incorporate the 2002 NSR reforms or had those revisions approved by EPA.

Identifying NSR Noncompliance Requires Long Investigations

The second major challenge EPA and state and local agencies face in ensuring compliance with NSR is that it is often difficult for regulators to identify noncompliance—that is, instances where owners did not obtain NSR permits before making major modifications to their generating units. According to several EPA officials we interviewed, identifying noncompliance can be challenging because unit owners—not regulatory agencies—have responsibility for determining whether they need an NSR permit. Most owners do not ultimately obtain NSR permits before making changes to their units, according to EPA officials we interviewed, because the owners determine that the changes fall under one of NSR's exclusions, such as routine maintenance, or because they offset emissions increases through netting. These unit owners are generally not required to notify EPA or state or local permitting agencies when they use these exclusions. Therefore, EPA would not review the owners' determinations unless (1) the owner proactively sought a permit and the state or local permitting agency determined that an NSR permit was required or (2) EPA initiated an investigation. In instances where a unit did not apply for and receive a permit as required, it can take EPA several years to identify the noncompliance and take corrective action. Moreover, under an EPA rule finalized in 2007, known as the "reasonable possibility recordkeeping" rule, a unit owner who determines that a change will not trigger NSR is not required to keep records of the change and its resulting emissions unless the owner believes there is a reasonable possibility that the change could result in a significant emissions increase, and other conditions are met.[22] According to one state official we interviewed, this rule may complicate efforts to identify noncompliance because EPA and state regulators generally have to retroactively determine whether an NSR permit should have been obtained for past activities, and without the benefit of company records, such a determination is difficult.

According to EPA and state officials we interviewed, state and local permitting agencies are generally not well positioned to identify noncompliance. State and local permitting agencies routinely inspect units, but officials told us these inspections focus on compliance with the terms of existing operating permits, not on whether an owner failed to obtain a permit. Several EPA and state officials told us that, given the complexity of most units, routine compliance inspections are not well suited to detect NSR violations, in part because it is difficult to distinguish work that might be considered a major modification from other work that is routine. According to one EPA official, to identify noncompliance with NSR, agency investigators need to identify what changes have already occurred; gather information on the nature of these changes; and determine whether NSR should have applied at the time the changes occurred, considering all possible exclusions and other factors. EPA officials we spoke with said that this process requires investigators to analyze information on historic emissions and a large volume of records on work conducted over the course of a unit's life. According to these and other EPA officials, such extensive review would not be possible during routine compliance inspections. Several state and EPA officials we spoke with also said that, given the complexity and case-by-case nature of NSR, state and local agencies generally do not have the resources—and in some cases expertise—to detect noncompliance. As result, several state officials we spoke with said they rely on EPA to identify instances of noncompliance with NSR.

EPA has therefore taken a lead role in enforcing NSR, beginning in the mid-1990s and continuing to the present. In 1996, EPA began targeting older, coal-fired generating units for compliance assessments and, on the basis of its investigations, alleged that several of the largest coal-fired electricity generating units in the country had violated NSR provisions by making major modifications without obtaining an NSR permit. In 1999 and early 2000, after receiving a number of cases from EPA, DOJ filed seven enforcement actions in U.S. federal courts in what is known as EPA's Coal-Fired Power Plant Enforcement Initiative. For their part, owners of units targeted by the NSR enforcement initiative contended that, among other things, their projects should have qualified for the routine maintenance exclusion. Nonetheless, almost all of these cases ultimately resulted in settlements mandating the installation of emissions controls and civil penalties. Since then, EPA and DOJ have continued this enforcement initiative and secured additional settlements for alleged noncompliance with NSR. According to EPA, steps to develop an NSR enforcement case include:

1. *Section 114 requests.* Under Section 114 of the Clean Air Act, EPA may obtain information from owners of generating units to determine whether violations have occurred. Such information includes detailed cost information on capital construction projects suspected to be NSR violations. According to EPA officials, collecting and reviewing such information can take several months to over a year.
2. *Settlement negotiations.* After reviewing generating units' records, EPA determines whether NSR violations have occurred. If EPA determines that the unit is not in compliance, it will notify owners of generating units and encourage the owner to install emissions controls. EPA initially tries to resolve noncompliance through a settlement.
3. *Referral.* If settlement negotiations are unsuccessful, EPA will determine whether enough evidence exists to refer the case to DOJ for potential litigation. DOJ then reviews the accumulated evidence and determines whether there is merit to file suit against the company. Before filing the case in court, DOJ generally discusses the matter with the owner in a further attempt to settle.

According to EPA and DOJ officials, EPA's investigations for NSR compliance, and subsequent enforcement actions, take a long time to conclude and involve substantial EPA resources. In instances where EPA's investigations have uncovered suspected violations, it can take years to litigate a case or bring it to conclusion through a settlement. Specifically, the 22 settlements resulting from EPA's enforcement initiative took, on average, 7 years to conclude. According to several industry representatives we interviewed, these efforts have also placed a large burden on owners and operators of generating units, given the amount of information required on past activities at the unit.

Available Data on NSR Compliance Are Not Complete but Suggest Substantial Noncompliance

Available data, while not complete, suggest that a substantial number of generating units have not complied with requirements to obtain NSR permits. Complete data on NSR compliance do not exist for two primary reasons. First,

EPA has not yet investigated all electricity generating units for compliance with requirements to obtain NSR permits. Second, NSR compliance is determined at a point in time, and EPA's interpretation of compliance has, in some cases, differed from that of federal courts. Nonetheless, EPA has investigated a majority of coal-fired generating units, and data from these investigations suggest that a substantial number of generating units have not complied.

Complete Data on Unit Compliance with NSR Are Not Available

From our review of relevant documentation and EPA-provided data, we identified two primary reasons why complete data on NSR compliance are not available. First, EPA has not yet investigated all generating units for NSR compliance, and second, available data do not provide a complete picture of compliance.

EPA has investigated most—but not all—coal-fired generating units for compliance with NSR at least once. According to our review of EPA-provided documents and data, EPA has investigated 831 generating units at least once since it began its Coal-Fired Power Plant Enforcement Initiative. These 831 units represent about 81 percent of all coal-fired units that generated electricity in 2010 and about 24 percent of all fossil fuel-fired units (those using coal, natural gas, or oil) that produced electricity in 2010.[23] Most natural gas units—as well as some smaller coal-fired units—have not been investigated by EPA. According to EPA officials we interviewed, the agency has focused most of its NSR compliance efforts on large, coal-fired units because they produce dramatically higher levels of harmful air emissions.

Data on units investigated by EPA are not conclusive because compliance is determined at a point in time; therefore, subsequent changes to the unit could affect its future compliance with NSR. NSR is required each time an existing generating unit undertakes a major modification. Thus, an owner of electricity generating unit that has obtained an NSR permit in the past—or was subject to an EPA investigation—is not exempt from the requirement to obtain an NSR permit for any future major modifications. [24] Moreover, allegations of noncompliance stemming from EPA's investigations do not necessarily mean that a violation has occurred, because in some cases federal courts have ultimately disagreed with EPA about the need for an NSR permit. Given these issues, it is difficult to provide a comprehensive assessment of NSR compliance at any given time.

EPA Has Found Noncompliance at a Majority of Units It Has Investigated

These limitations notwithstanding, data on generating units investigated by EPA shed some light on overall compliance with NSR among electricity generating units. Among the 831 generating units it has investigated through its Coal-Fired Power Plant Enforcement Initiative, EPA has alleged NSR noncompliance at a substantial number. Specifically, EPA has alleged noncompliance—documented through notices of violation, complaints filed in court, or settlements—at 467 generating units, around 56 percent of the units it has investigated.[25] These 467 units investigated by EPA represent approximately 45 percent of coal-fired units that produced electricity in 2010, and about 14 percent of all fossil fuel-fired units that produced electricity in 2010.[26]

According to EPA, the Coal-Fired Power Plant Enforcement Initiative is perhaps the most comprehensive and coordinated enforcement effort under the Clean Air Act to date. The initiative has led to 22 settlements covering a total of 263 units, or approximately 32 percent of the units EPA has investigated. According to our analysis of EPA data, the settlements will require affected unit owners to install and operate emissions controls costing an estimated $12.8 billion in total and levy civil penalties totaling around $80 million. Some companies are also required to fund environmentally beneficial projects, such as restoring watersheds and forests in national parks. These settlements are projected to reduce sulfur dioxide emissions by more than 1.8 million tons annually and nitrogen oxides emissions by about 596,000 tons annually.[27] In some cases, EPA reached companywide settlements in which companies agreed to put emissions controls on units constituting most of their production capacity. Two of the largest settlements—with American Electric Power and the Tennessee Valley Authority—represent 105 units, around 40 percent of the total, and about $8.6 billion in control costs, or around two-thirds of the total. A senior Department of Justice official we interviewed said that, in addition to the 22 concluded settlements, 7 additional NSR cases are in various stages of litigation. See appendix III for more details on EPA's concluded NSR settlements.

CONCLUSIONS

Ensuring compliance with NSR presents significant challenges for EPA, and state and local permitting agencies. Unit owners—not regulators— are responsible for determining when an NSR permit is needed, and, because of NSR's complex requirements, these determinations are not always straightforward. The complexity and case-by-case nature of NSR requirements also mean that NSR violations are difficult to detect, once they occur. State and local agencies generally do not have the resources to identify NSR violations and therefore rely on EPA to enforce NSR. Although available data on NSR compliance are not conclusive, the substantial number of generating units EPA investigations have allegedly found to be noncompliant suggests that many generating units have not obtained NSR permits as required. Addressing NSR's complexity and improving compliance could reduce the need for long and resource-intensive enforcement actions and more effectively protect air quality by averting emissions before they occur. Yet EPA's ability to simplify NSR or develop comprehensive, nationwide guidance is limited for several reasons, including the case-by-case nature of NSR applicability, ongoing litigation, and the variation in NSR requirements across states.

Nonetheless, EPA has an opportunity to improve its efforts by collecting more comprehensive NSR permitting data. Several EPA regional offices maintain some information on the NSR permits issued by the state and local permitting agencies in their regions, but this information is in different formats and not compiled by EPA into a complete and centralized source of information on NSR permits issued nationwide, as recommended by the National Research Council in 2006. More complete information on NSR permitting would help EPA and external parties gauge the extent to which fossil fuel generating units have obtained NSR permits and help inform enforcement efforts that have already found widespread alleged noncompliance.

In cases where unit owners apply for permits before making physical or operational changes that would result in a significant net increase of emissions, EPA plays an important role because it has an opportunity to comment on every draft NSR permit under consideration by state and local permitting agencies and to influence decisions about the appropriate level of pollution control, among others. A key benefit of EPA's involvement in the permitting process is that the agency can review and comment on permits issued in different geographic areas and assess various aspects of draft permits, including the level of emissions control required. Because emissions controls can cost owners and operators of generating units hundreds of millions of dollars, EPA's review of the required level of emissions control is critically important. Although EPA and headquarters staff devote resources to commenting on draft permits, EPA does not track whether state and local permitting agencies incorporate the agency's comments. Without such information, EPA cannot fully assess the extent to which state and local agencies incorporate its comments in NSR permits or the extent to which emissions control requirements imposed by state and local permitting agencies reflect suggestions made by EPA in its comments.

RECOMMENDATIONS FOR EXECUTIVE ACTION

To help improve EPA's implementation of NSR, we recommend that the EPA Administrator direct the entities responsible for implementing and enforcing NSR—specifically, the Office of Enforcement and Compliance Assurance, Office of Air Quality Planning and Standards, and EPA regions—to take the following two actions:

- Working with EPA regions and state and local permitting agencies, consider ways to develop a centralized source of information on NSR permits issued to fossil fuel electricity generating units, and
- Using appropriate methods, such as sampling or periodic assessments, develop a process for evaluating the effects of its comments on draft NSR permits.

AGENCY COMMENTS AND OUR EVALUATION

We provided a draft of this report to the Department of Energy, the Department of Justice, and Environmental Protection Agency (EPA). The Department of Energy said they had no comments on the report's findings and recommendations. The Department of Justice provided technical comments, which we incorporated as appropriate. In its written comments, EPA agreed with the importance of having good systems for tracking and compiling information to efficiently and effectively administer its programs, while enhancing accountability and transparency, but disagreed with the need for the actions called for in our recommendations.

Regarding our first recommendation that EPA work with state and local permitting authorities to consider ways to develop a centralized source of information on permits issued to electric generating units, EPA said that it believes it has a number of permit tracking mechanisms in place, and raised four concerns about our recommendation. First, EPA said that it has maintained a centralized permit information database for many years—the RACT/BACT/LAER Clearinghouse, which is capable of capturing and sharing information on NSR permits that have been issued. However, EPA acknowledged that this database is incomplete—including about half of issued NSR permits—primarily because, in some areas, state and local agencies are not required to enter information about the permits they issue. Nonetheless, EPA said it is taking steps to improve participation. We continue to believe that comprehensive permitting data would enable EPA, Congress, and other interested parties to better understand the scope and impact of NSR.

Second, EPA said that its regional offices track NSR permitting by the states in their jurisdiction and that the agency believes it is most appropriate for the regional offices, rather than headquarters, to be responsible for this information. However, our work found that the tracking of NSR permits by EPA's regional office varied in completeness. For example, of the four regions we included in our sample, one region had a robust system for tracking issued NSR permits, and one had no system at all. EPA also said that its regional offices provide oversight of state and local agencies and that an EPA-wide compilation of permit data would be redundant, add costs, and provide little benefit to its oversight function. We continue to believe that a centralized source of complete information on NSR permits would enhance EPA's oversight of state and local permitting agencies and help ensure consistency across regions. EPA headquarters could build on the ongoing efforts of some regional offices and develop more complete data using a simple, low-cost system. For example, we found that two regional offices use a spreadsheet to compile and maintain basic data on permits issued by state and local agencies. Additionally, we believe that any costs incurred in developing more comprehensive data should be considered relative to the benefits that could accrue from having better information on the universe of permitted facilities including, as noted by the National Research Council, the ability to assess the impact of policy changes.

Third, EPA said that a centralized database of all NSR permits would not help most members of the public because most members of the public are interested in permits issued to specific facilities rather than the entire universe of all permits issued. Our report focused on the importance of more complete data to enhance programwide oversight of NSR permitting and targeting of enforcement efforts. More complete data could potentially assist the public and other interested parties in understanding the extent of NSR permitting for individual facilities, but this was not the basis of our findings and recommendations. We continue to believe that a centralized source of permitting data is important for EPA's oversight of state and local permitting agencies and to enhance its enforcement efforts.

Fourth, EPA questioned the value of more comprehensive information in targeting noncompliance with requirements to obtain permits. Specifically, EPA said that identifying noncompliance involves targeting facilities that should have obtained permits but did not and that information on facilities that have obtained permits would not assist in these efforts. Moreover, EPA said that getting data on noncompliant sources is time- and resource-intensive. We continue to believe that compiling complete information on facilities that have obtained permits could help identify facilities that have not obtained permits and enhance targeting of these facilities for potential noncompliance. We also believe that understanding which facilities have obtained permits as required could decrease these time and resource demands because the agency would have a better starting point for identifying noncompliance.

Regarding our second recommendation that EPA develop a process for evaluating the effect of its comments on issued permits, the agency said that its regional offices already do so and described the interactions between these offices and state and local agencies during the permitting process. EPA also said that its regional offices already conduct oversight of state and local permitting agencies, including whether these agencies adequately address EPA's comments on draft permits. We acknowledge these efforts in the report and believe that, as part of its overall oversight of nationwide permitting efforts, EPA headquarters could benefit from a broader and more comprehensive assessment of the extent to which its comments on draft permits were adequately considered and incorporated. Because the terms of issued permits can result in the installation of pollution controls that cost hundreds of millions of dollars, it is important to conduct higher level review of issued permits to identify variability in the terms of issued permits across geographic areas. We therefore continue to believe that implementing this recommendation would enhance oversight of NSR permitting nationwide and that EPA has an opportunity to build on the information already collected through the oversight activities of its regional offices.

EPA also provided technical comments that we incorporated as appropriate.

Sincerely yours,

David C. Trimble
Director, Natural Resources and Environment

Frank Rusco
Director, Natural Resources and Environment

APPENDIX I. SCOPE AND METHODOLOGY

To assess what information the Environmental Protection Agency (EPA) maintains on New Source Review (NSR) permits issued for fossil fuel electricity generating units, we gathered information from EPA and selected states on the status of their NSR permitting programs and efforts to collect and maintain permitting data. We selected a nonprobability sample of nine states on the basis of (1) the number of older electricity generating units in the state; (2) the quantity of electricity generated by such units in those states; (3) the volume of sulfur dioxide, nitrogen oxides, and carbon dioxide emitted by units in those states; and (4) the region in which the generating unit was located.[28] We obtained these data from the Ventyx Velocity Suite EV Market-Ops database, a proprietary database containing consolidated energy and emissions data from EPA, the Energy Information Administration (EIA), and other sources. To assess the reliability of the Ventyx data, we reviewed documentation provided by Ventyx and tested key variables to verify their accuracy and determined the Ventyx data to be sufficiently reliable for our purposes. The nine states we selected were Alabama, Georgia, Indiana, Kentucky, Missouri, New York, North Carolina, Ohio, and Pennsylvania. To assess how permitting information is collected and used, we reviewed relevant documentation from these nine states and from EPA. We also interviewed permitting officials from these nine states, the four EPA regional offices that oversee these states, EPA's Office of Air and Radiation, its Office of Inspector General, and its Office of Enforcement and Compliance Assurance.

In three of the states, some localities are responsible for NSR permitting; we also spoke with officials at two of those localities, which we selected on the basis of the number of older units in their jurisdictions.

To examine what challenges, if any, EPA, state, and local agencies face in ensuring compliance by electricity generating units with requirements to obtain NSR permits, we reviewed relevant provisions of the Clean Air Act and NSR regulations; guidance and other information on implementing NSR maintained by EPA; and literature on NSR from government agencies, academic and research institutions, environmental organizations, and industry groups. We also interviewed knowledgeable officials and stakeholders from these agencies and institutions, as well as officials from the selected states and localities.

To review what available data show about compliance with requirements to obtain NSR permits, we reviewed information published by EPA on the estimated rate of noncompliance by industrial sectors. We also reviewed information on EPA's enforcement activities maintained by enforcement officials in EPA's Office of Enforcement and Compliance Assurance, including (1) data on notices of violation sent to owners of generating units alleging noncompliance with NSR; (2) lawsuits filed in court for alleged NSR violations; and (3) information on the settlements concluded by EPA and the Department of Justice with owners of generating units, which ended or prevented lawsuits alleging noncompliance. To assess the reliability of the EPA-provided data, we interviewed knowledgeable agency officials and tested key variables to verify their accuracy. We determined these data to be sufficiently reliable for the purposes of our analysis. We also interviewed knowledgeable enforcement and compliance officials from EPA's headquarters Office of Enforcement and Compliance Assurance and four regional offices.

We conducted this performance audit from April 2011 to June 2012, in accordance with generally accepted government auditing standards. Those standards require that we plan and perform the audit to obtain sufficient, appropriate evidence to provide a reasonable basis for our findings and conclusions based on our audit objectives. We believe that the evidence obtained provides a reasonable basis for our findings and conclusions based on our audit objectives.

Appendix II. Comparison of Available Databases of NSR Permits for Electricity Generating Units

Table 1. Comparison of Available NSR Permit Databases

Database	Main purpose	Limitations
EPA control technology clearinghouse	To enable state permitting agencies, EPA regional offices, and regulated sources to assess EPA's determinations on past NSR applications with respect to required emissions control technologies	By EPA's estimates, captures about 50 percent of permits
State databases or permit records	To assist states in keeping records of their own permitting activity	Do not consolidate permits for regionwide or nationwide assessment; state systems to track permits vary widely
Regional databases	To inform regulated sources and the public on the status of permitting activity	Data limited to regional jurisdiction; completeness of data varies by region
Office of Air Quality Planning and Standards greenhouse gas permitting database	To assist EPA's headquarters permitting division in tracking NSR permits related to greenhouse gases following 2010 greenhouse gas rule	Tracks permits issued since 2011, after greenhouse gas rule took effect; may exclude NSR permits for some activities, such as coal handling, that affect particulate matter but not greenhouse gases
Office of Air Quality Planning and Standards comment letter database	To help EPA ensure consistency in its comments on draft permits and to meet comment deadlines	Does not track whether draft permits are finalized, or their terms
Region 7 coal-fired utility database	To track permits issued to coal-fired generating units from 2000 to 2010	Database only covers newly constructed coal-fired units for a limited time period
Region 4 combustion turbine database	To track permits issued to combustion turbine units from the early 1990s onward	Database only covers combustion turbine units, most commonly natural-gas-fired units, and is no longer being updated

Source: GAO analysis.

Appendix III. Concluded NSR Settlements

Table 2. Concluded NSR Settlements Involving Electricity Generating Units

Dollar in millions

Company	Date settlement lodged	Number of units involved	Civil penalties	Estimated control costs	Environmentally beneficial projects	Emissions reductions (tons per year)	
						Sulfur dioxide	Nitrogen oxides
Tampa Electric Company	2/19/2000	10	$3.5	$1,000	$10.5[a]	70,000	53,000
Public Service Enterprise Group	1/24/2002 and 11/30/2006	4	7.4	337	9.25	36,194	18,807
Alcoa, Inc.	3/27/2003	3	1.5	330	2.5	52,900	15,480
Virginia Electric and Power Company	4/17/2003	20	5.3	1,200	13.9	176,500	60,400
Wisconsin Electric Power Company	4/29/2003	23	3.2	600	22.5[b]	72,300	32,600
Southern Indiana Gas and Electric Company	6/6/2003	3	0.6	30	2.5	6,400	4,200
South Carolina Public Service Authority	3/16/2004	12	2	400	4.5	37,500	29,500
Illinois Power Company and Dynegy Midwest Generation	3/7/2005	10	9	500	15	39,500[c]	14,800[d]
Ohio Edison	3/18/2005 and 8/11/2009	29	8.5	1,100	25	171,500	31,050
Alabama Power	4/25/2006	2	0.1	200	4.9	22,790	4,940
Minnkota Power Cooperative	4/26/2006	3	0.85	100	5	23,561	9,458
Nevada Power	6/13/2007	4	0.3	60	0.4	N/A	2,300
Eastern Kentucky Power Cooperative	7/2/2007	6	0.75	600	47	54,000	8,000
American Electric Power	10/9/2007	46	15	4,600	60	654,000	159,000

Table 2. (Continued)

Company	Date settlement lodged	Number of units involved	Civil penalties	Estimated control costs	Environmentally beneficial projects	Emissions reductions (tons per year)	
						Sulfur dioxide	Nitrogen oxides
Salt River Project	8/12/2008	2	0.95	400	4	14,303	6,789
Kentucky Utilities	2/3/2009	1	1.4	135	3	28,500	3,000
Duke Energy	12/22/2009	4	1.75	85	6.25	35,000	N/A
Westar Energy	1/25/2010	3	3	500	6	60,000	18,600
American Municipal Power	5/18/2010	4	0.85	0	15	30,600	3,160
Hoosier Energy	7/23/2010	4	0.95	275[e]	5	19,827	1,845
Northern Indiana Public Service Company	1/13/2011	11	3.5	600	9.5	46,000	18,000
Tennessee Valley Authority	4/14/2011	59	10	4,000[f]	350	225,757	115,977
Total		**263**	**$80**	**$12,777**	**$589**	**1,837,632**	**596,106**

Source: GAO analysis of EPA data.

[a] EPA estimated the costs of environmentally beneficial projects at $10 to $11 million.

[b] EPA estimated the costs of environmentally beneficial projects at $20 to $25 million.

[c] EPA estimated reductions in sulfur dioxide emissions of 39,500 tons per year compared with 2003 emissions and 283,000 tons per year compared with 1999 emissions.

[d] EPA estimated reductions in nitrogen oxides emissions of 14,800 tons per year compared with 2003 emissions and 58,200 tons per year compared with 1999 emissions.

[e] EPA estimated control costs at $250 to $300 million.

[f] EPA estimated control costs at $3 to $5 billion.

End Notes

[1] Fossil fuel units are responsible for nearly all emissions of carbon dioxide, sulfur dioxide, and nitrogen oxides from the electricity generating sector.

[2] EPA has set national ambient air quality standards for six pollutants, termed "criteria" pollutants: carbon monoxide, lead, nitrogen oxides, ozone, particulate matter, and sulfur oxides.

[3] This report focuses solely on fossil fuel electricity generating units and major New Source Review at those units, although NSR also applies to certain other major stationary sources of air pollution, such as other industrial facilities. NSR also applies to minor modifications at major stationary sources, as well as to new and modified minor sources (known as minor NSR).

[4] In areas that meet EPA's national ambient air quality standards—known as attainment areas—the NSR permitting process includes a review for "prevention of significant deterioration" to ensure that the emissions will not exceed maximum allowable increases for three criteria pollutants: nitrogen oxides, sulfur dioxide, and particulate matter. In areas that do not meet the standards—known as nonattainment areas—the permitting process is known as "nonattainment NSR." Throughout this report, the term NSR includes both types of review.

[5] What constitutes a significant net increase in emissions depends on whether the generating unit is located in an attainment or nonattainment area and the type of regulated pollutant being emitted. For example, in all areas, a 100-ton-per-year increase is significant for carbon monoxide, while a 40-ton-per-year increase is significant for nitrogen oxides or sulfur dioxide. In attainment areas, increases of noncriteria regulated pollutants can also be significant.

[6] GAO, *Air Emissions and Electricity Generation at U.S. Power Plants,* GAO-12-545R (Washington, D.C.: Apr. 18, 2012). Analysis is limited to generating units that (1) listed a fossil fuel as a primary fuel; (2) generated electricity in 2010; and (3) had a net summer capacity greater than 25 megawatts, making them subject to EPA's emissions monitoring and reporting requirements.

[7] These 1,485 older units produced 45 percent of the electricity from fossil fuel units in 2010. The remainder was produced by 1,958 newer units (those that began operating after 1978).

[8] Other revisions to NSR regulations were proposed in 2006 and 2007 but were never finalized, or, if finalized, EPA has stayed their effective date, is considering whether changes to the regulation are necessary, or both. Similarly, a revision finalized in December 2008 has been stayed.

[9] "Indian country" includes all land within the limits of an Indian reservation under the jurisdiction of the United States government, all dependent Indian communities within the borders of the United States, and all Indian allotments, the Indian titles to which have not been extinguished.

[10] Some states, however, have laws preventing agencies from issuing more stringent regulations.

[11] The Cross-State Air Pollution Rule and the National Emissions Standards for Hazardous Air Pollutants from Coal-and Oil-Fired Electric Utility Steam Generating Units are currently being challenged in court. A federal appeals court has stayed the effective date of the Cross-State Air Pollution Rule while it hears the case. GAO is currently undertaking a study on the effects of these two rules on reliability. The study is expected to be issued in 2012.

[12] Our review was limited to the NSR permits issued to electricity generating units; state and local permitting agencies may also maintain information on NSR permits issued to other stationary sources.

[13] Regional offices also maintain information on the NSR permits issued by their office, such as to generating units in Indian Country, although officials said that regional offices rarely issue such permits.

[14] National Research Council, *New Source Review for Stationary Sources of Air Pollution* (Washington, D.C.:2006). In its report, the Council found that no centralized source existed for information on NSR permits issued nationwide. It estimated—using data provided to EPA by state and local permitting authorities that were preliminary, unpublished, and not subjected to review—that from 1997 to 1999 there had been several NSR permits issued for major modifications to existing sources in the combined electricity, gas, and sanitary services sector: 38 for carbon monoxide, 30 for particulate matter, 46 for nitrogen oxides, and zero for sulfur dioxide.

[15] The RACT/BACT/LAER Clearinghouse is a database maintained by EPA's Office of Air and Radiation, which contains information provided by state and local permitting agencies on air pollution control technologies, including technologies required in source-specific permits. The database contains information on over 5,000 air pollution control permit determinations submitted by permitting agencies for several U.S. territories and all 50 states.

[16] *See* 42 U.S.C. § 7503(d). According to EPA officials, in many cases, EPA regional offices also have planning agreements with states stipulating that the state must enter data from all NSR permits into the database. Some regional offices specify a certain time frame by which states must enter the data. These officials said that, as a general matter, the planning agreements are tied to grant funding received by states, and EPA could conceivably withhold funding if a state fails to meet a condition of the agreement. However, with declining budgets and resources at state government levels, officials said that entering data is not always a priority for states, or they may choose to defer entering data until they believe they have a permit that is noteworthy.

[17] For example, EPA maintains a database aimed at tracking its reviews of draft state-issued permits; the database does not, however, contain information on final permits or their terms.

[18] EPA regulations require state and local permitting agencies to give the public at least 30 days to comment on draft NSR permits.

[19] In states that have adopted their own NSR regulations, these NSR rules may differ from EPA's, although they must be no less stringent than EPA's regulations.

[20] The portion of the generating unit's emissions post-project that could have been accommodated by the generating unit at a certain time before the proposed change, and that are also unrelated to the project, including any increased utilization due to demand growth, can be excluded from the calculation of emissions increases.

[21] In September 2006, EPA proposed a rule clarifying how emissions decreases from a project may be included in the calculation to determine if a significant emissions increase will result. EPA did not finalize the rule.

[22] This 2007 final rule was a clarification of a 2002 NSR reform regulation that the D.C. Circuit Court remanded to EPA in 2005. EPA is currently considering whether changes to the regulation are necessary and a litigating a case about the rule.

[23] Analysis is based on a universe of 3,443 electricity generating units in the U.S. that (1) listed a fossil fuel as a primary fuel, (2) had a net summer capacity greater than 25 megawatts, and (3) produced electricity in 2010.

[24] Although units must undergo NSR review for major modifications, some of the settlement agreements EPA has reached with electricity generating units include a provision precluding EPA, in certain circumstances, from suing the owner for making a major modification and not undergoing NSR.

[25] In the settlements, companies generally denied the alleged violations and maintained their compliance with the Clean Air Act but agreed to settle to avoid the costs and uncertainties of litigation and improve the environment.

[26] Analysis is based on a universe of 3,443 electricity generating units that (1) listed a fossil fuel as a primary fuel, (2) had a net summer capacity greater than 25 megawatts, and (3) produced electricity in 2010.

[27] These reductions are to be phased in over an agreed-upon time frame, often 10 years.

[28] Because we used a nonprobability sample, the results we obtained from electricity generating units in these states are not generalizeable to such units in all states; nonetheless, these results did help us understand how NSR is implemented and enforced in different states.

INDEX

A

B

C

D

G

H

I

J

L

M

N

O

P

S

T

U

V

W

Y